Database Theory and Application

Database Theory and Application

Rex Porbasas Flejoles

ARCLER
PRESS

www.arclerpress.com

Database Theory and Application

Rex Porbasas Flejoles

Arcler Press
2010 Winston Park Drive,
2nd Floor
Oakville, ON L6H 5R7
Canada
www.arclerpress.com
Tel: 001-289-291-7705
001-905-616-2116
Fax: 001-289-291-7601
Email: orders@arclereducation.com

ISBN: 978-1-77361-221-8 (Hardcover)

ABOUT THE AUTHOR

Rex Porbasas Flejoles

Rex P. Flejoles is an IT Faculty member of a state university in the Philippines. He is involved in software development, research, and data analysis. He graduated Bachelor of Science in Computer Engineering in 2002. Moreover, he received Master of Science in Computer Science degree in 2006. In addition, he completed the academic requirements for PhD in Science Education major in Mathematics Education. He has completed a number of research studies and served as adviser and statistician to various undergraduate research studies.

TABLE OF CONTENTS

LIST OF FIGURES

PREFACE

Databases are fundamental elements of information systems. In spite of the fact that database ideas, innovation, and designs have been produced and combined in the most recent decades, numerous viewpoints are responsible for the innovative development and transformation. In connection, composing a book on this traditional but persistently advancing field is an incredible test.

Key Components

This book covers various areas of databases, which makes it a good material for learning the fundamentals and advanced topics. It offers an adjusted perspective of ideas, dialects/languages, and models, with solid reference to current innovation on business database management systems (DBMSs).

This book is made of ten principle parts. This book is intended to clarify the standards of database administration and to acquire two fundamental abilities: how to operate a database (including the preparation of program to access the database) and how to outline its design structure. These are the essential parts of outlining and controlling a database that are required to make compelling utilization of database innovation.

Author

CHAPTER 1

INTRODUCTION

Accumulation, association, and capacity of data and information are real considerations in numerous human transactions and in every computer-based system. Bank adjustments, phone numbers in registries, stock citations, accessibility of credit cards reserves, and registers of students enlisted in a school with their grades in exams are some cases where sorts of data and information are crucial in the completion of pertinent transactions. Computer-based frameworks guarantee that these data and information are properly kept, regularly updated, and securely made available to corresponding clients from a remote area or within the computer system where the data and information are kept. Envision, for instance, questions about the accessibility of assets on Visas, asked through certain gadgets accessible in millions of organizations (regardless of whether inns, shops, or organizations), which permit buying at any place on the planet to be charged to the credit card proprietors.

This book is about the administration of data and information using electronic information systems. This portrays the ideas important about data and information in a computer system, the languages for update and retrieval as well as the specific structures for the administration of the database systems. In the first chapters, the ideas of data frameworks and databases are presented, with the goal that we would then be able to manage the significant qualities of frameworks for overseeing databases.

DATA AND INFORMATION

In the completion of any transaction, whether by an individual or an organization of any size, the accessibility of data and information as well as and the ability to oversee them are necessary. In this way, every organization has an information system, which deals with the data that are important in activities of the organization. The presence of the data framework is somewhat autonomous to a degree to which it is mechanized. Take note that information systems existed some time before the creation and across the board selection of computers. For instance, bank records and constituent rolls have been set up for a few centuries already. Mostly, just a part of the information system of an organization is electronic. The spread of computing innovation into all human movements, normal for the eighties and of the nineties, creates a nonstop development in computerization of information systems.

In the most straightforward of human activities, data and information are recorded and traded as per the regular methods average of the action itself such as composed or talked language, drawings/pictures, charts, and numbers. In few activities, an expressed record would not even exist, the data being pretty much precisely remembered. Regardless, we can state that as activities have progressively moved toward becoming systematized, proper types of association and codification have been found (ElMasri, 1994).

In computer-based frameworks, the idea of recording and arranging is taken as far as possible. The data are recorded by certain methods, which need to be processed so as to give corresponding information. Similarly as with numerous essential ideas, it is hard to give an exact meaning of certain data as well as the difference between the data and information. Generally, we can state that data alone are of lesser value. However, once these data are appropriately processed, information will be made available which present us better insight of the world.

For instance, the string John Smith and the number 25775, composed on a sheet of the paper, are two different data that are of lesser value when not being processed. But, in an instance where and inquiry is made, for example, 'Who is the leader of the exploration division and what is his phone number?', at that point it is conceivable to process those data and present corresponding information such as a person named John

Smith is the leader of the exploration division and that his phone number is 25775.

Having presented the idea of data and information along this line, we can proceed to that of the database as this is the primary theme of this material. Different definitions are conceivable. One way of defining it is as follows: a database is a collection of gathered data that is used to provide data of interest to an information system. In this book, we will principally consider a significantly morc limited definition, identified with a particular innovation, talked about in the following segment. In numerous applications, information is inherently substantially steadier than the (manual or programmed) methods that work upon it. Coming back to the aforesaid illustration, we can see that the information identifying with bank applications does have a structure in which intents and purposes remained unaltered for a considerable length of time, while the systems that follow up on it differ persistently, as each client can promptly confirm (Ramakrishnan, 1997).

Besides, when one method is substituted for another, the new strategy "acquires" the information of the old one, with suitable changes. This characteristic of security drives us to express that information constitutes to be an "asset" for the association that oversees it, a huge legacy to exploit and secure.

DATABASE MANAGEMENT SYSTEMS THEORY

Consideration regarding information has portrayed computer-based applications a long ago, yet programming systems devoted particularly to the administration of information have been in presence in the late sixties and few applications still did not utilize them. Without an intended system for databases, information administration is performed by methods for customary programming languages such as C and Fortran, or with the most recent like C++ and Java. Numerous applications are developed in 'Cobol', a programming language specifically designed for 'business applications' that means applications essentially composed for information management.

The regular way to deal with information administration exploits the presence of records to store information forever. A record takes into consideration the capacity and seeking of information, but gives just basic mechanisms for accessibility and sharing. With this approach, the methodology written in a programming language is totally self-sufficient. Every one characterizes and uses at least one or more "private" records. Information of conceivable interest to more than one program is reproduced the same number of times as there are programs that utilize it, with clear excess and the likelihood of irregularity. Databases were made, generally, to conquer this kind of problem.

We are presently in the position of characterizing a database management system (DBMS) as a product system ready to keep and maintain information that are vast, shared, and tireless, as well as to guarantee their unwavering quality and protection. Like any programming item, a DBMS must be proficient and powerful. A database is a collection of data and information taken care of by a DBMS.

Let us take a remark on each of the above recorded attributes of the DBMS.

- Databases could be expansive, as in they can contain great billions of bytes and are unquestionably bigger than the accessible primary memory. Correspondingly, a DBMS must oversee information in auxiliary memory. Clearly "little" databases can exist, however the frameworks must be ready to keep and maintain data and information without being constrained by dimensions, aside from the physical ones of the current gadget available.
- Databases are shared, as in different applications and clients have the capacity to access information of common interest. It is imperative to specify that in this way anomalies are minimized if not totally eliminated, since redundancies are avoided, and consequently, the likelihood of anomalies is lessened. On the one hand, if more than one duplicate of similar data exist, it is conceivable that they are not indistinguishable. On the other hand, if each datum is put away just once, at that point the issue of irregularity is disposed of. Keeping in mind the end goal to ensure shared access to information by numerous clients working all the time, the DBMS makes utilization of an exceptional system called concurrency control.
- Databases are diligent. It means that they have a life expectancy

that is not constrained to single executions of the projects that utilize them. Then again, take note that the information overseen by a program in primary memory has an existence that starts and closes with the execution of the program. In this way, such information is definitely not persistent.

- DBMS guarantee unwavering quality. It means that the system is capable of protecting the substance of the database (or if nothing else to permit its reproduction) on the off chance that the equipment or programming fails. To satisfy this prerequisite, DBMS gives particular capacities for reinforcement and recovery.
- DBMS guarantee information security. Every client, who is authenticated based on certain level of authority, determined through the login credentials, will be given corresponding access to the system.
- DBMS considers productivity that is the ability to do operations utilizing a suitable measure of assets (time and space) for every client. This trademark depends on the procedures utilized as a part of the usage of the DBMS and on how good the item has been outlined. It ought to be focused on that DBMS gives a colossal mix of components that require numerous assets and frequently put substantial necessities on the assets given by the working condition.
- Finally DBMS increments adequacy, referring to the limit of the database system to make the activities of its clients beneficial, in each sense. This is obviously a non-specific definition and does not compare to a particular work, given that a DBMS gives different administrations and capacities to distinctive clients. The undertaking of designing a database and the applications that utilize it point basically to ensure the great, general adequacy of the system.

It is notable that the administration of substantial and steady accumulations of data and information is additionally conceivable by using less modern than DBMSs, starting with the documents shown in every single working system. Documents were acquainted to manage "local" information to a particular technique or application (Silberschatz, 1996).

DBMSs were imagined and made to expand the elements of the document systems, allowing shared access to similar information by more than one client and application. Furthermore, DBMSs ensure coordination among the numerous different administrations. Plainly, DBMSs in their turn utilize records for the capacity of information.

MODELS OF DATA

A data model is a blend of builds used to sort out data. Every data display gives organizing components, like the sort constructors of programming languages, which permit the meaning of new information sorts based on constructors connected to predefined, rudimentary sorts. For instance, Pascal permits the development of types by methods for cluster, record, set and document constructors.

The relational data model gives the relation constructor that makes it conceivable to sort out data in an accumulation of records with a settled structure. A relation is frequently spoken to as a table, in which lines correspond to particular records and its sections refer to the fields of the record; the request of lines and sections is insignificant. For illustration, information about college course offerings in certain degree program with the respective tutor may be presented in two tables, namely TEACHING and PROSPECTUS, as depicted in Figure 1. As should be obvious, a relational database includes numerous relations.

TEACHING	
Course	**Tutor**
Databases	Smith
Networks	Jones
Languages	Robinson

PROSPECTUS	
Degree Programme	**Subject**

Information Systems	Databases
Information Systems	Networks
Information Systems	Languages
Electrical Engineering	Databases
Electrical Engineering	Networks

Figure 1 - Relational Database – A Simple Example.

(Source: Silberschatz, A. K. (Database System Concepts. New York.)

The relational model was proposed in an exploration setting toward the start of the seventies. It showed up to be available in genuine systems in the mid-eighties and is presently the broadest. Other than the relational model, three different sorts of model have been characterized.

- The hierarchical model in light of the utilization of tree-like structures (and chains of importance, subsequently the name), characterized amid the principal period of advancement of DBMSs (in the sixties) yet at the same time utilized as a part of numerous systems, predominantly for progression reasons.
- The network model (otherwise called the CODASYL model, after the Conference on Data Systems Languages that gave it an exact definition), in light of the utilization of charts, created in the mid-seventies.
- The object-oriented database model, created in the eighties to overcome a few constraints of the relational model, which stretches out to databases the worldview of object-oriented programming.

The data models recorded above are for the most part accessible in business DBMSs. They are called consistent. To underline the way that the structures utilized for these models, albeit conceptual, mirror a specific association (tree, chart, table, or object). Other data models known as semantic models, have been acquainted to portray information in a way free of the coherent model; be that as it may, these are not accessible in business DBMSs. Their name originates from the actuality that they have a tendency to portray ideas of this present reality, as opposed to the information required for their portrayal. These models are utilized as a part of the preparatory period of the database configuration process, to examine the application in the best conceivable route, without implementation 'pollutions'.

Schema and Instances

In a database there is a section that is invariant in time, called the design of the database, made up of the schema, and a section that progresses with time, called the instance of the database, composed of the real values. In the case of Figure 1, the relations have a settled structure; the TEACHING relation has two columns (called attributes), which allude to course and tutor. The pattern of a connection comprises of its first segment, which is the name of the connection, trailed by the names of its attributes. Below is an illustration.

TEACHING (Course, Tutor)

Databases	Smith
Networks	Jones
Languages	Robinson

The columns of the table fluctuate with time, and relate to the courses presently offered and the respective tutors. During the life-span of the database, tutors and courses are inserted, deleted, or updated correspondingly. The outline is adjusted starting with one year then onto the next. The occasion of a connection is comprised of an accumulation of lines, which changes with time.

We could likewise say that the schema is the intensional segment and the instance is the extensional segment. These definitions will be developed and examined more in Chapter 2.

Abstraction Levels

The ideas of the model and schema depicted above can be further developed, considering different measurements in the display of information. In specific, an institutionalized design for DBMSs has been proposed. It is partitioned into three levels, referred to separately as physical, logical, and view; for each level there is a schema.

The physical schema is a depiction of the entire database by means of the legitimate model received by the DBMSs (that is, one of the models portrayed prior: relational, hierarchical, and object-oriented).

The logical schema depicts the usage of the consistent construction by methods for physical capacity structures. For instance, a connection can be physically sorted out as a successive record, or a hash document, or a consecutive document with files.

A view schema is the portrayal of a part of the database by means of the legitimate model. A view level can offer an alternate association of the information so as to mirror the perspective of a specific client or gathering of clients. In this way it is conceivable to connect different outer diagrams with a solitary sensible composition: each of the outer constructions will give a particular perspective of the database (or a subset).

In the vast majority of the present frameworks, the view level is not unequivocally present; be that as it may, it is conceivable to characterize inferred relations (called views). Taking into consideration the database in Figure 1 for instance, a student in Electrical Engineering degree program could be intrigued just in the courses incorporated in the said curriculum. These data are available in the Electrical Engineering instance, which can be retrieved from the PROSPECTUS table considering just the lines that allude to the Electrical Engineering degree program. Instruments for access approval can be related with outside diagrams, so as to control the entrance of clients to the database. A client could be approved to control just the information depicted by means of his or her view schema.

PROSPECTUS		
Degree Programme	**Subject**	**Year**
Electrical Engineering	Databases	4
Electrical Engineering	Networks	4

Figure 2 – Relational View.

(Source Silberschatz, A. K. Database System Concepts. New York.:)

Data Independence

The multilevel engineering portrayed in the past segment ensures information freedom, a noteworthy property of DBMSs. As a rule, this property permits clients and application projects to allude to information at an abnormal state of deliberation, disregarding usage subtle elements. In a more absolutely way, data independence presents two angles—physical and logical. Physical data independence permits cooperation with the DBMS independently of the physical parts of the information. For instance, it is conceivable to adjust the association of the records that actualize the relations or the physical distribution of records to the capacity gadgets without impacting the abnormal state depictions of the information and projects that information utilization.

While, logical data independence ensures that the communication with the outside level of the database is free of the legitimate level. For instance, it is conceivable to include another outer pattern as per requests of another client or to adjust a conceptual schema without modifying the existing schema, hence hide the physical association of the information. In the meantime, it is conceivable to change the coherent level, keeping up unaltered the outer outline that is important to certain client (given that its definition as far as the consistent structures is balanced) (O'Neil P. , 1994).

Underline that entrance to a database happens just through the view schema (which can concur with the logical level). It is the DBMS that deciphers the operations as far as the basic levels. The multilevel design is subsequently the crucial component through which DBMSs accomplish data independence.

LANGUAGES AND CLIENTS

DBMSs are perplexing systems that offer an assortment of languages for the administration of information and include, in their life cycle, an assortment of clients or users.

Languages of the Database

Taking note of the qualification among patterns and examples presented above, we may recognize database languages comparably.

- The Data Definition Language (DDL) is utilized to define the structure of the database including such as the attributes and their corresponding data types.
- The Data Manipulation Language (DML) is utilized for retrieving and updating the database.

Note that few languages, such as, SQL, which will be discussed in detail in Chapter 4, offer the components of both the DDL and DML in an incorporated shape. Access to the data can be indicated in different ways.

- Through intuitive printed languages, for example, SQL.
- Through charges like the intelligent ones installed in customary programming languages, such as C++, Cobol, and Fortran. These are called have languages since they "have" charges written in the databases' language.
- Through charges like the intuitive ones installed in ad hoc advanced languages, often with particular components (for instance, for diagrams, complex printouts, or structures and menus).
- Through well-disposed interfaces that permit the detailing of queries without the utilization of a textual, literal language. These too, contrast extraordinarily from one system to another and are constantly advancing.

An extensive number of projects for information section, administration, and printing have a typical structure. Thus, the presence of advanced language in addition to benevolent interfaces, extensively disentangle the generation of applications, decreasing advancement time and expenses.

Users and Creators

Different classes of individuals can connect to a database through a DBMS. Below are the most vital.

- The database administrator (DBA) is the person in charge of the plan, control and organization of the database. The DBA has the assignment of intervening among the different prerequisites, frequently clashes, communicated by the clients, ensuring unified control over the database. Specifically, he or she is in charge of ensuring administrations, guaranteeing the unwavering quality of the system, and dealing with the approvals for access to the database.

- The application planners and software engineers characterize and make programs that connect to the database. They utilize the DML in conjunction to other relevant tools to prepare corresponding interfaces for the users of the application.
- The clients or users utilize the database for their respective exercises and activities. They can be classified into two—end clients and casual clients.
    - End clients, who utilize the program with predefined exercises and activities, with couple of special cases known and considered in advance.
    - Casual clients, ready to utilize the intelligent language to access the database, figuring queries (or updates) of different sorts. They can be experts in the language they utilize and associate as often as possible with the database. Note that the expression "easy-going" implies that the database queries are most certainly not predefined.

PROS AND CONS OF DBMSS

We close this section by outlining the basic qualities of databases and DBMSs, and their favourable circumstances and inconveniences.

- DBMSs enable database (collection of data) to be considered as a typical asset of an association, accessible to all its approved individuals.
- The database gives an institutionalized and exact model of that piece of the reality important to the association, usable in already in-place applications and with the fundamental augmentations in future applications.
- With the utilization of a DBMS, unified control of the information is conceivable, which can be enhanced by types of institutionalization and can profit by an 'economy of scale'.
- Sharing permits the decrease of repetition and irregularity.

- Data autonomy, the central characteristic for DBMSs, favours the advancement of applications that are more adaptable as well as effectively modifiable.

The utilization of DBMSs likewise conveys some negative angles, like:

- DBMSs are costly items, complex and very unique in relation to numerous other programming instruments. Their presentation along these lines speaks to an impressive speculation, both direct (the cost of the item) and circuitous (the securing of the essential equipment and programming assets, application relocation, and workforce).
- DBMSs give, in institutionalized shape, an entire arrangement of administrations, which fundamentally conveys costs. In the situations where some of these administrations are most certainly not required, it is hard to extricate the administrations really required from other administrations - services, and this can create waste.

Taking everything into account, we can state that circumstances can exist in which the selection of a DBMS can be awkward and difficult. A few applications with one or only a couple of clients can be cheaper to be delivered offline. But lately, DBMS innovation has advanced impressively, bringing about progressive and more effective and solid systems, on more far reaching and less costly design, so expanding the comfort of creating applications with a DBMS.

CHAPTER 2

RELATIONAL DATABASES

Most present database system depends on the relational model, with the expectation of giving a premise to data independence. The foundation of the relational model as an accepted standard was somewhat moderate, because of its high level of abstraction. Productive arrangements were not quickly found for relational structures, which were not quite the same as those being used around then.

Despite the fact that the principal models of relational system had been made in the mid-seventies, the main relational system showed up available in the mid-eighties, gaining a noteworthy offer of it just in the mid-eighties. This section shows the basic parts of the actual model, and the way it is utilized to sort out information. After a concise discussion of the different logical models, there is an indication of how the idea of a connection can be utilized in dealing with the data in a database. At that point, we quickly talk about the strategies for the portrayal of fragmented data and we look at limitations, which permit the particular of extra properties that must be fulfilled by the database.

The introduction of the relational model is presented in two parts. The first part is committed to the standards of query operations on relational databases, while the second part is about SQL—the language utilized as a part of genuine system for the definition, update and query on relational databases.

THE RELATIONAL MODEL THEORY

Logical Models

The relational model depends on two ideas, connection (or relationship) and table, which vary in their inclination yet are much related. The idea of connection is formal, as it originates from science, specifically from set hypothesis, while the idea of table is basic and instinctive. Their concurrent nearness is presumably the real explanation behind the immense achievement of the relational model. Indeed, tables offer a characteristic seeing even to end clients who run over them in numerous settings other than in databases. Then again, the accessibility of an unmistakable and basic formalization has enabled the improvement of a hypothesis to help the model, with extremely fascinating outcomes.

The relational model fulfils the prerequisite of data independence, as examined in Chapter 1. Clients and software engineers make reference just to relations (legitimate or outside level), which are then executed by methods for proper physical structures. To access information, it is most certainly not important to think about the physical structures themselves. The relational model was proposed toward the finish of the sixties keeping in mind the end goal that is to give a higher level of data independence than the system and progressive models. These included reference to the basic structure, by methods for the utilization of pointers and the physical request of information(Codd E. F., A Relational Model for Large Shared Data Banks, 1970).

Relations and Tables

Reviewing some concept in arithmetic courses, when given two type of sets, D1 and D2, the Cartesian result of D1 and D2 , in images D1xD2, is the arrangement of requested sets (v1,v2) , with the end goal that v1 is a component of D1 and v2 is a component of D2 . For instance, if we take into consideration A = {1, 2, 4} and B = {a, b}, the Cartesian item AxB is the arrangement of all the conceivable matches in which the primary component has a place with A and the second to B. Since A has three (3) components and B has two (2), we have six (6) elements in pairs:

{(1, a), (1, b), (2, a), (2, b), (4, a), (4, b)}

In arithmetic, a connection on the sets D1 and D2 (called areas of the connection) is a subset of D1xD2. Given the sets A and B over a case connection on A and B, a possible subset is this set {(1, a), (1, b), (4, b)}.

The above definition does not demonstrate whether the sets we are thinking about might be limited or not, and along these lines incorporate the likelihood of unending sets (furthermore, along these lines of unending relations). Practically speaking, since our databases must be put away in computer systems of limited size, the relations are essentially limited. At the same time, it can be valuable for the spaces to have an unbounded size (with the goal that it is constantly conceivable to accept the presence of a value or esteem not present in the database). Accordingly, we will expect that our databases are made up of limited relations on potentially endless areas.

Relations can be spoken to graphically in table frame. The two tables appeared in Figure 3 portray the Cartesian item AxB and the numerical connection on A and B showed previously. The meanings of Cartesian item and connection allude to two sets, yet can be summed up regarding the quantity of sets. Given $n>0$ sets D1, D2, ... Dn not really particular, the Cartesian result of D1, D2, ... Dn, spoke to by D1xD2x...Dn, is comprised of the arrangement of the n-tuples v1,v2, ... vn , with the end goal that vi has a place with Di for $1<i<n$. A scientific connection on the areas D1, D2, ... Dn is a subset of the Cartesian item D1xD2x...Dn.

1	a
1	b
2	a
2	b
4	a
4	b

1	a
1	b
4	b

Figure 3 – Table Form of the Cartesian Item

(Source: Atzeni, P. D. Relational Database Theory)

The number n of the segments of the Cartesian item (and along these lines of each n-tuple) is known as the level of the Cartesian item and of the connection. The quantity of components (that is, of n-tuples) of the connection is, as normal in set hypothesis, the cardinality of the connection. Figure 4a demonstrates the unthinkable portrayal of a Cartesian item on the areas C={x, y}, D={a, b, c} and E={3, 5}, with degree 3. Figure 4b demonstrates a connection among CxDxE with a degree of 3 and cardinality 6 (Codd E. F., 1982).

(a)

x	a	3
x	a	5
x	b	3
x	b	5
x	c	3
x	c	5
y	a	3
y	a	5
y	b	3
y	b	5
y	c	3
y	c	5

(b)

x	a	3
x	a	5
x	c	5
y	a	3
y	c	3
y	c	5

Figure 4 - A ternary Cartesian Product and Relation

(Source: Atzeni, P. D. Relational Database Theory)

Relations (and the comparing tables) can be utilized to speak to information for any application. For instance, the connection in Figure 5 contains information relating to the results of soccer matches.

Real Madrid	Liverpool	3	1
Liverpool	Milan	2	0
Real Madrid	Roma	1	2
Roma	Milan	0	1

Figure 5 – Soccer Matches Relation Results

(Source: Atzeni, P. D. Relational Database Theory)

It is characterized with reference to two areas Integer and String, each of which shows up twice. The connection is in actuality a subset of the Cartesian item:

'String × String × Integer × Integer'

Relations with Properties

We can mention different objective facts about relations and their properties. As indicated by the definition, a scientific connection is a set of requested n-tuples (v_1, v_2, ... v_n). With reference to the utilization that we make of relations for sorting out information in our database, we can state that every n-tuple contains different things of information associated with each other or sets that join them. For illustration, the main n-tuple of the connection in Figure 5 builds up a connection between the qualities Real Madrid, Liverpool, 3, 1 to demonstrate the result of the match held from Real Madrid versus Liverpool is 3 to 1. We can at that point recall that a connection is a set and consequently:

- There is no characterized arrangement between the n-tuples. In the tables that speak to them, there is clearly an "introduction" arrangement, but is insignificant since two tables with similar lines yet in various requests speak to the same connection.
- The n-tuples of a connection are particular one from the other. Since among the components of a set there cannot be two indistinguishable components. In this way, a table can speak to a connection just if its lines are unique in relation to each other.

In the meantime, every n-tuple has a request: the i-th estimation of every one originates from the i-th area. This is fundamental for understanding the importance of the information in the connection. If we somehow happened to swap the third and fourth parts around in the connection in Figure 5, it would totally change the significance of our connection. In that, the consequences of the matches would be altered. This happens in light of the fact that each of the two whole integer and string shows up twice in the connection, and the two events are recognized on the premise of their positions. The primary appearance of the string alludes to the home group and the second to the meeting group, correspondingly, the two events of the integer (Tsichritzis, 1982).

This request among the areas of a connection really relates to an unacceptable normal for the idea of connection as characterized in arithmetic as to the likelihood of sorting out and utilizing information.

Without a doubt, in software engineering there is a propensity to incline toward non-positional documentations to positional ones. The previous one enables reference to the fields of a record by means of representative names. While, the last one alludes to similar fields through their request and ought to be utilized just when the request compares to an inborn element, for instance, in numerical examination issues in which clusters offer an undeniable and direct portrayal of vectors and matrices. The information that we wish to sort out in the relations with our databases has a structure that is fundamentally the same as that of records, a connection is generously an arrangement of homogeneous records that is characterized on the same fields. Consequently, we present a non-positional documentation by the same names with the areas in a connection, alluded to as qualities, which depict the “parts” played by the areas. For instance, for the connection concerning the matches, we can utilize names like HomeTeam, VisitingTeam, HomeGoals, and VisitorGoals. In the unthinkable portrayal, we utilize the characteristics as segment headings as presented in Figure 6. Given the need of recognizing the segments unambiguously, the characteristics of a connection (and along these lines the segment headings) must be not the same as each other.

HomeTeam	VisitingTeam	HomeGoals	VisitorGoals
Real Madrid Liverpool	Liverpool	3	1
Real Madrid	Milan	2	0
Roma	Roma	1	2
	Milan	0	1

Figure 6 – A Relation.

(Source: Ullman, J. D. Principles of Database and Knowledge Base Systems)

By adjusting the meaning of connection with the presentation of qualities, before we give the definition, we can actually see that the request of characteristics (and of the sections in the unthinkable portrayal) is insignificant. It is not any more important to talk about first area, second space, and so on. It is adequate to allude to the traits or attributes. Figure 6 demonstrates another forbidden portrayal of the connection in Figure 7 with the traits and in this way the sections in an alternate request (after the American style in which the home group is appeared after the meeting group).

VisitingTeam	HomeTeam	VisitorGoals	HomeGoals
Liverpool	Real Madrid	1	3
Milan	Liverpool	0	2
Roma	Real Madrid	2	1
Milan	Roma	1	0

VisitingTeam	HomeTeam	VisitorGoals	HomeGoals
Liverpool	Real Madrid	1	3
Milan	Liverpool	0	2
Roma	Real Madrid	2	1
Milan	Roma	1	0

Figure 7 – Another Representation of Figure 6

(Source: Ullman, J. D. Principles of Database and Knowledge Base Systems.)

To formalize the ideas, let us set up the correspondence between characteristics and areas by methods for a capacity DOM: $X \rightarrow D$, which partners with each characteristic $A \in X$ an area $DOM(A) \in D$. At that point, let us say that a tuple on an arrangement of traits X is a capacity t, which partners with each quality $A \in X$ an estimation of the area DOM(A). In this way, we can give the new meaning of connection: a connection on X is an arrangement of tuples on X. The difference between this definition and the customary one of set hypothesis dwells just in the meaning of tuple. In a scientific connection, we have n-tuples whose components are recognized by position. Though in the new definition, the components are recognized by the characteristics that are by a non-positional method. Starting now and into the foreseeable future, we will utilize the expression "connection" to allude to the new non-positional definition.

Give us a chance to present a valuable documentation that we will utilize much of the time later on. On the off chance that t is the tuple on X and $A \in X$, at that point t[A] (or t.A) shows the estimation of t on A. For instance, if t is the main tuple of the connection in Figure 7, we can state that:

't[VisitingTeam] = Liverpool'

A similar documentation is additionally utilized for sets of qualities, in which case it indicates a tuple: '*t[VisitingTeam, VisitorGoals]*' is a tuple on two attributes.

THE STRUCTURE OF THE RELATIONAL MODEL - DATABASES AND RELATIONS

As we talked and just observed, a connection can be utilized to sort out significant information for an application. In any case, a solitary connection is not typically adequate for this reason. A database is by large comprised of a few relations, in which tuples contain regular values (or esteems) where this is fundamental to set up correspondences. Let us investigate this idea by taking into consideration the database in Figure 8.

Exams	Student	Grade	Course
	276545	C	01
	276545	B	04
	937653	B	01
	200768	B	04

COURSES	Code	Tide	Tutor
	01	Physics	Grant
	03	Chemistry	Beale
	04	Chemistry	Clark

Figure 8 – Database in a Relational Model

(Source Ullman, J. D. Principles of Database and Knowledge Base Systems)

- The main connection contains data about students, namely: student number (RegNum), surname (Surname), first name (FirstName), and date of birth (BirthDate).
- The third connection contains data on courses: Code, Title, and Tutor.
- The second connection contains data about exams: Student, Grade, and Course. This connection makes reference to the information contained in the other two tables: through RegNum in STUDENTS table, while Code in COURSES table.

The database in Figure 8 demonstrates one of the basic attributes of the relational model, which is regularly communicated by saying that it is 'value-based'. The references between information in various relations are spoken to by methods for the estimations of the spaces that show up in the tuples. Rather than the network and hierarchical models that were characterized before the relational model, speak to references unequivocally by methods of pointers and this is the reason why they are called 'pointer-based' models. Since in this book we do not have an itemized introduction of these models, we quickly remark here the crucial attributes of a basic model with pointers. Figure 9 demonstrates an indistinguishable database that is the same as that in Figure 8, wherein we have utilized pointers rather than value-based references (the enlistment quantities of the students and course)(Maier D. , 1983).

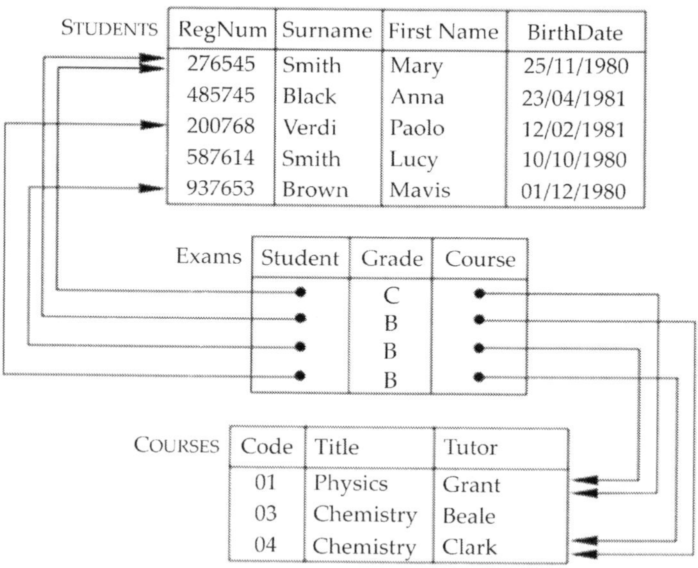

Figure 9 – Pointers Used in a Database

(Source: Ullman, J. D. Principles of Database and Knowledge Base Systems)

Contrasted with a pointer-based model, the relational model has a few focal points.

- The legitimate portrayal of information (containing just values) does not make any reference to the physical one, which can likewise shift with time. Accordingly, the relational model permits physical data dependency.

- It speaks to just what is important from the perspective of the application (the client). The pointers are extra, identified with implementation perspectives. In the models with pointers, the application software engineer needs to allude to information that is not huge for the application.
- Given that the greater part of the data is contained in the qualities, it is generally easy to exchange the information starting with one setting then onto the next (for instance, on the off chance that it is important to move the database starting with one system then onto the next). With pointers, the operation is more difficult in light of the fact that the pointers imply, which is nearby to every system (local), thus cannot be sent out or exported in a direct manner.

We should note the fact that, even in a relational database, on the physical layer, the information can be spoken to by methods that require the utilization of pointers. The distinction, concerning pointer-based models, is that these pointers are not obvious at the consistent level. We can recap now the meanings of the relational model, with a barely any points of interest, recognizing the levels of the patterns from those of the cases.

- A connection diagram comprises of an image, called name of the connection R and a set of (names of) properties X={A1, A2, ... An}, the entire usually shown as R(X). A space is related with each trait or attribute.
- A database construction comprises of an arrangement of connection outlines with various names: R= {R1(X1), R2(X2), ... Rn(Xn)}
- A connection occasion (or just connection) on a mapping is a set of r tuples in X.
- A database occasion (or just database) on a mapping R = {R1(X1), R2(X2), ... Rn(Xn)} is an arrangement of relations r = {r1, r2, … rn}, where each ri is $1<i<n$, is a connection on the construction Ri(Xi).

To give a case, we can state that the database outline in Figure 7 is:

R = {STUDENTS (RegNurn, Surname, FirstName, BirthDate),

EXAMS (Student, Grade, Course),

COURSES (Code, Title, Tutor)}

For comfort, we abridge the conventions that we embrace from now on (and that we have effectively utilized as a part of the definitions and in the specific cases and examples).

- The qualities (when they do not utilize critical names from the application perspective) will be shown by letters close to the start of the letter set in capitals, perhaps with a superscript or a subscript: A, B, C, A'.
- Sets of characteristics will be shown by letters close to the finish of the letters in order in capital letters: X, Y, Z, X'. A set whose parts we wish to feature will be indicated by the juxtaposition of the names of the characteristics themselves; we will compose X=ABC instead of X={A, B, C}. Comparatively, the union of sets will be indicated by the juxtaposition of the comparing names; we will compose XY as opposed to XuY. Consolidating the two traditions, we will compose XA rather of Xu{A}.
- For the connection names we will utilize the letter(s) R (and S) in capital such as: R1, S, S'. For the connection occasions, we will utilize indistinguishable images from the comparing connection names, yet in bring down case.

To outline the principal ideas of the relational model, let us examine two more illustrations.

In the first place we will take note of how relations on a solitary property are acceptable. This bodes well in databases comprising of a few relations, in which a connection on a solitary characteristic contains values that show up as estimations of a characteristic of another connection. Let us consider the database with STUDENTS connection as illustrated in Figure 8, it is conceivable to utilize another connection on a single characteristic: RegNum, to show the students who are also workers (by methods for the important RegNum, which must show up in the STUDENTS connection) (see Figure 10).

STUDENTS	RegNum	Surname	FirstName	BirthDate
	276545	Smith	Mary	25/11/1980
	485745	Black	Anna	23/04/1981
	200768	Verdi	Paolo	12/02/1981
	587614	Smith	Lucy	10/10/1980
	937653	Brown	Mavis	01/12/1980

WORKERS
RegNum
276545
485745
937653

Figure 10 – A Single Attribute Relation

(Source: http://pld.cs.luc.edu/database/)

Let us take a gander at a more perplexing case, which indicates how the relational model permits the portrayal of data with an itemized structure. In Figure 11, three receipts from a restaurant are listed. They have a structure that (aside from the pre-printed states in strong writing) incorporates some important data (receipt number, date, and total amount) and a variable number of lines, each alluding to a homogeneous gathering of dishes (quantity, description, and cost). Since our relations have a settled structure, it is not possible and impractical to speak to the gathering of receipts with a solitary connection. It would not be conceivable to speak of various dishes that are not known prior in time (Ullman J. D., 1988).

Da Mario		
Receipt No: 1357		
Date: 5/5/92		
3	covers	3.00
2	hors d' oeuvre	5.00
3	first course	9.00
2	steak	12.00
	Total:	29.00

Da Mario		
Receipt No: 2334		
Date: 4/7/92		
2	covers	2.00
2	hors d' oeuvre	2.50
2	first course	6.00
2	bream	15.00
2	coffee	2.00
	Total:	27.50

Da Mario		
Receipt No: 3007		
Date: 4/8/92		
2	covers	3.00
2	hors d' oeuvre	6.00
3	first course	8.00
1	bream	7.50
1	salad	3.00
2	coffee	2.00
	Total:	29.50

Figure 11 – Different Receipts

(Source: Maier, D. The Theory of Relational Databases)

We can speak to similar information by methods of two relations, as appeared in Figure 11. The connection RECEIPTS contains the information indicated as it was once in each receipt (receipt number, date, and total amount) and the connection DETAILS contains the different lines of each different receipt (quantity, description, and), related with a similar receipt through the receipt number.

RECEIPTS		
Number	**Date**	**Total**
1357	5/5/92	29.00
2334	4/7/92	27.50
3007	4/8/92	29.50

DETAILS			
Number	Quantity	Description	Cost

1357	3	covers	3.00
1357	2	hors d'oeuvre	5.00
1357	3	first course	9.00
1357	2	steak	12.00
2334	2	covers	2.00
2334	2	hors d'oeuvre	2.50
2334	2	first course	6.00
2334	2	bream	15.00
2334	2	coffee	2.00
3007	2	covers	3.00
3007	2	hors d'oeuvre	6.00
3007	3	first course	8.00
3007	1	bream	7.50
3007	1	salad	3.00
3007	2	coffee	2.00

Figure 12 – Database for Figure 11

(Source: Maier, D. The Theory of Relational Databases)

The important things for the corresponding database of Figure 11 to represent the receipts accurately, two conditions must be fulfilled:

- It is not important to monitor the request in which the lines show up on each receipt. Actually, since no request is characterized among the tuples of a connection, the tuples in DETAILS are not in any settled request.
- No copy lines show up on any receipt (which could occur for the situation of various requests for similar dishes with a similar amount).

The two issues can be settled by including a characteristic, which demonstrates the position of the line of the receipt. Thus, it is constantly conceivable to recreate impeccably the substance of the considerable number of receipts as depicted in Figure 12.

RECEIPTS		
Number	**Date**	**Total**
1357	5/5/92	29.00
2334	4/7/92	27.50
3007	4/8/92	29.50

DETAILS				
Number	**Line**	**Quantity**	**Description**	**Cost**
1357	1	3	covers	3.00
1357	2	2	hors d'oeuvre	5.00
1357	3	3	first course	9.00
1357	4	2	steak	12.00
2334	1	2	covers	2.00
2334	2	2	hors d'oeuvre	2.50
2334	3	2	first course	6.00
2334	4	2	bream	15.00
2334	5	2	coffee	2.00
3007	1	2	covers	3.00
3007	2	2	hors d'oeuvre	6.00
3007	3	3	first course	8.00
3007	4	1	bream	7.50
3007	5	1	salad	3.00
3007	6	2	coffee	2.00

Figure 13 – Database for Figure 11 (Another One)

(Source: Maier, D. The Theory of Relational Databases.)

Incomplete Data and Invalid Values

The specific structure of this relational model, as previously mentioned, is extremely straightforward and effective. In the meantime, it forces a certain level of inflexibility, in that the data must be spoken to by methods for homogeneous tuples of information. Specifically, in any connection we can represent just tuples comparing to the composition of the connection. Actually, in numerous cases, the accessible information would not compare precisely to the picked format.

For instance, consider the connection outlined below.

"PERSONS(Surname, FirstName, Address, Telephone)"

The value of the Telephone attribute might not be accessible for all the tuples. It is significant to note, that it would not be right to utilize a value of the area or domain to represent to a nonappearance and absence of data as this would create perplexity. For this situation, assuming the phone numbers are indicated by numbers, we could, for instance utilize zero to demonstrate the nonappearance of the noteworthy value. But, this decision is inadmissible, for two reasons.

First, it requires the presence of a value of the area or domain that has never utilized before for noteworthy values. On account of phone numbers, zero is unmistakably discernible, yet in different cases there exists no accessible incentive for this reason. For instance, in a property that speaks to the date of birth and that utilizes a domain a sort Date accurately characterized. There are no "unused" components that would hence be usable to indicate nonattendance of data. Second, the utilization of domain or area values can create perplexity. The refinement between really huge values and imaginary ones ('placeholders') is covered up. Consequently, the projects that approach the database must consider recognition among them (and monitoring which are the imaginary values for each situation).

So it can be represented simply, yet in the meantime advantageously, the non-availability of qualities, the idea of connection is normally stretched out to incorporate the likelihood that a tuple can accept, on each trait either an estimation of the space as observed up to now or an uncommon value. The invalid value signifies a nonattendance of data and is an extra value. In the unthinkable portrayals, we will use for the invalid value the NULL symbol, as illustrated in Figure 14, which manages the addresses of government workplaces in the cities of Italy. It can be noted that regional towns have nearby government workplaces but the other towns do not. If we refer to the table in the figure, we could take note of how as a result, the three invalid values that show up in it are allocated for various reasons.

City	GovernmentAddress
Rome	Via Quattro Novembre
Florence	NULL
Tivoli	NULL
Prato	NULL

Figure 14 –Invalid Values

(Source: Atzeni, P. D. Relational Database Theory. California.)

- Florence is a district town and in that capacity should positively have a nearby government office. Right now we do not have the idea about its address. The invalid value is utilized on the grounds that the genuine esteem cannot be recorded in the database. Thus, we say that it is an obscure value.
- Tivoli is not a province town and along these lines has no neighbourhood government office. Hence, the Government Address property can have no value or data for this tuple. Here, the null or the invalid value signifies the inapplicability of the characteristic, or in other words, the non-presence of the value—the value or data is non-existent.
- The province of Prato has been set up as of late and we do not know whether the neighbourhood government office has not been opened yet nor do we know its address (regardless of whether officially operational or arranged). As a result, we do not know whether the value exists; on the off chance that it exists, we do not realize what it is. In reality, we wind up in a circumstance that compares to the consistent disjunction (the 'or') of the two going before—the value is either non-existent or obscure. This kind of invalid values and data is typically called null, since it let us know literally nothing. The value may or may not exist. In the event that it exists, we do not know what it is.

In relational database frameworks, no speculation is made about the significance of invalid value. In this way, practically speaking, we wind up in the third circumstance that is null value.

For a further reflection on invalid values, consider now the database in Figure 15, which is characterized on an indistinguishable database design from Figure 8. The null value on the date of birth in the first tuple of the STUDENTS connection is pretty much permissible. In that, one can acknowledge that the data is definitely not basic in this unique situation. But for a case that the value for the enrolment number or the course code is null makes significant issues, since these qualities, as we have talked about with reference to Figure 8 are utilized to build up correspondences between tuples of various relations. Now, the presence of null values in the EXAMS connection really makes the data unusable. As illustrated in the second tuple, only the Grade has a value while the other two attributes contain null values, gives no valuable information. In this way, the presence null or invalid values in a connection can in a few cases produce questions with regards to the genuine importance and personality of this tuples. The last two (2) tuples of the COURSES connection can be unique or can actually concur. Consequently, the need for keeping a reasonable control over the utilization of null values in our relations is obvious. Just certain connection examples ought to be conceded. All in all, when a connection is characterized, it is conceivable to indicate that null values are permissible just for a few qualities and not for all (Paredaens, The Structure of the Relational Database Model, 1989).

STUDENTS

RegNum	Surname	FirstName	BirthDate
276545	Smith	Mary	NULL
NULL	Black	Anna	23/04/1972
NULL	Verdi	Paolo	12/02/1972

EXAMS

Student	Grade	Course
276545	C	01
NULL	B	NULL
200768	A	NULL

COURSES

Code	Title	Tutor
01	Physics	Grant
03	Chemistry	NULL
NULL	Chemistry	Clark

Figure 15 – Database with Invalid Values

(Source: Atzeni, P. D. Relational Database Theory. California.)

INTEGRITY REQUIREMENTS

The structures of the relational model enable us to sort out the data of considerable interest to our own applications. In the most cases, however, it is not genuine that each set of tuples in the mapping speaks to data that is right for the application. We have just examined the issue quickly as regarding the connections of null values. Presently, we will take a gander at the issue in more noteworthy detail. First, alluding to relations without null values. For illustration, consider the database in Figure 16 and note the different circumstances that ought not to happen.

STUDENTS

RegNum	Surname	FirstName	BirthDate
200768	Verdi	Paolo	12/02/1981
937653	Smith	Lucy	10/10/1980
937653	Brown	Mavis	01/12/1980

EXAMS

Student	Grade	Honours	Course
200768	K		05
937653	B	honours	01
937653	A	honours	04
276545	C		01

COURSES

Code	Title	Tutor
01	Physics	Grant
03	Chemistry	Beale
04	Chemistry	Clark

Figure 16 – Incorrect Information in a Database

(Source: Maier, D. The Theory of Relational Databases. Potomac, Md.)

With the end goal of this exercise, we will expect that the highest Grade is A in which "honours" is granted, while the lowest Grade is F.

- In the principal tuple of the EXAMS connection, we have a K value for the Grade. Supposedly, this is not acceptable since the Grade must be within A to F only.
- In the second tuple of the EXAMS connection, a distinction is granted for a B value for the Grade just like the one having the A value.
- The last two tuples of the STUDENTS connection contain data on two different students but with the same RegNum. This is a mistake of assigning the same student number to two different

students.

- The fourth tuple of the EXAMS connection contains record of a certain students who is not found in the STUDENTS table. This is likewise an inadmissible circumstance given that the student number furnishes us with data just as a connection to the comparing tuple of the STUDENTS connection. In addition, the first tuple shows a course code with the value of 05 that does not exist in the COURSES connection.

In a database, it is basic to maintain a strategic distance from circumstances, like those simply depicted. For this reason, the idea of uprightness limitation was presented, as a property that must be fulfilled by all the right database occurrences. Each requirement must be viewed as a predicate, which relates the value, genuine or false with each occurrence. Generally, we connect an accumulation of limitations with a database pattern and we consider remedy (or legitimate) the occurrences that fulfil each of the requirements. In each of the four cases presented above, an imperative would forbid the undesirable circumstance. It is conceivable to arrange the requirements as indicated by the components of the database that are engaged with it. There are two classes, with the first having some specific sub-cases.

- A limitation is intra-relational if its fulfilment is characterized with respect to a single connection of the database. The initial three cases above depict the intra-relational limitations. The meaning of the requirement considers the tuples (or even the qualities) independently from each other.
    - A tuple requirement is a limitation that can be assessed on each tuple freely from the others. The imperatives identifying with the initial two cases fall into this classification.
    - As a still more particular case, a limitation characterized with reference to single value (as in the principal case in which, for the quality Grade, just value among A and F are permitted) is known as a value imperative or domain/ space limitation, given that it forces a confinement on the area of the trait.

- A limitation is inter-relational on the off chance that it includes more than one connection. This is found in the fourth case, in which the unsuitable circumstance can be denied by asking for that a RegNum shows up in the EXAMS connection just in the event that it shows up in the STUDENTS connection.

In the accompanying segments we will inspect tuple limitations: key imperatives that are the most critical intra-relational requirements and referential imperatives that are the most essential between relational imperatives.

Tuple Imperatives

As we have stated, tuple imperatives express different conditions on the values of each particular tuple, freely of different tuples.

A conceivable language structure for these limitations allows the meaning of Boolean articulations (and that is, with some connectives such as AND, OR, and NOT) with atoms that think about values of traits (or arithmetical articulations utilizing estimations of properties). The abused imperatives in the initial two cases are portrayed by the accompanying articulations.

(Grade ³ A) AND (Grade £ F)

(NOT (Honours = ‘honours’)) OR (Grade = A)

Specifically, the second limitation shows that distinctions are permissible for a Grade with a value of A (telling that either there is no distinctions, or the grade is equivalent to A, or both). The principal imperative is in reality a domain or an area imperative, given that it includes a solitary trait. The definition we have given likewise concedes more mind boggling articulations, given that they are characterized on the estimations of single tuples. For instance, a connection for the pattern:

PAYMENTS (Date, Amount, Deductions, Net)

it is conceivable to characterize the imperative that forces the condition for the Net value as:

Net = Amount - Deductions

Keys

In this segment we will talk about key imperatives that are without a doubt the most critical of the relational model. We could even go far and state that without them, the model itself would have no sense. Let us illustrate it with an example. For the connection in Figure 17, the values of the different tuples of the RegNum characteristic are not quite the same as each other. The value in the RegNum unambiguously recognizes the students. The general concept of the student number itself was presented numerous years back, a long time before the presentation of databases, unequivocally keeping in mind the end goal to have a basic and viable instrument for alluding to the students in an unambiguous way. Essentially, in the connection there are no sets of tuples with values exactly the same for the three properties, namely: Surname, FirstName, and BirthDate. These snippets of data also distinguish every individual unambiguously.

Different arrangements of properties likewise recognize unambiguously the tuples of the connection in Figure 17. For instance, the match RegNum and DegreeProg, given that RegNum is adequate all alone.

RegNum	Surname	FirstName	BirthDate	DegreeProg
284328	Smith	Luigi	29/04/59	Computing
296328	Smith	John	29/04/59	Computing
587614	Smith	Lucy	01/05/61	Engineering
934856	Black	Lucy	01/05/61	FineArt
965536	Black	Lucy	05/03/58	FineArt

Figure 17 – Keys Illustration

(Source: Maier, D. The Theory of Relational Databases. Potomac, Md.)

Instinctively, a key is an arrangement of credits used to distinguish unambiguously the tuples in a connection. More unequivocally:

- an arrangement of traits K is a super-key for a connection r if r does not contain two unmistakable tuples t1 and t2 with t1[K]=t2[K]; and

- an arrangement of properties K is a key for which r if K is an insignificant super-key.

For such a case in Figure 17:

- The set RegNum is a super-key. It is additionally a negligible super-key, given that it contains a sole trait and along these lines we can reason that RegNum is a key.
- The set consisting of the Surname, FirstName, and BirthDate is a super-key. None of its subsets is a super-key. As illustrated above, there are two equivalent tuples (first and second) on Surname and BirthDate. Also, there are equivalent tuples (fourth and fifth) in terms of Surname and FirstName. Likewise, two equivalent tuples (third and fourth) as to FirstName and BirthDate. Hence, the set containing the Surname, FirstName, and BirthDate is another key.
- The set consisting of RegNum and DegreeProg is a super-key as we have seen. It is not an insignificant super-key, in light of the fact that one of its legitimate subsets, RegNum is itself a negligible super-key, and in this manner RegNum, DegreeProg is not a key.
- The set containing the FirstName and DegreeProg is not a super-key on the ground that there are two tuples in the connection. As seen in the figure above, equivalent tuples existed with similar values in the FirstName and DegreeProg.

Keeping in mind the end goal to talk about the subject in more profundity, let us look at another connection as illustrated in Figure 18. It contains no match of tuples that concede to both Surname and DegreeProg. Along these lines, for this connection, the set consisting of the Surname and DegreeProg is a super-key. Since there are tuples that concede to Surname (first and second) and on DegreeProg (second and fourth), this set is a negligible super-key and along these lines a key. Presently, in this connection, Surnames and DegreeProg distinguish the tuples unambiguously and would we be able to state this is valid.

RegNum	Surname	FirstName	BirthDate	DegreeProg
296328	Smith	John	29/04/59	Computing
587614	Smith	Lucy	01/05/61	Engineering
934856	Black	Lucy	01/05/61	FineArt
965536	Black	Lucy	05/03/58	Engineering

Figure 18 – Another key Illustration

(Source: Teorey, T. J. Design of Database Structures.)

Thus, we can state that the set containing the Surname and DegreeProg is 'by possibility' a key for the connection in Figure 18 while we are occupied with the keys comparing to trustworthiness requirements fulfilled by all the lawful relations on a specific blueprint. When characterizing a mapping, we connect with it imperatives that portray properties in this present reality, for which data are held in our database.

The limitations are characterized at the outline level, with reference to all the occasions that must fulfill every one of the limitations. A right occurrence would then be able to fulfill different requirements past those characterized in the pattern. For instance, with a composition:

STUDENTS(RegNum, Surname, FirstName, BirthDate, DegreeProg)

are related to the requirements that force as keys the two arrangements of characteristics talked above:

RegNum

Surname, FirstName, BirthDate

Both the relations in Figure 17 and Figure 18 fulfil both the limitations. The second likewise fulfills ('by shot', as we said) the requirement that says that Surname, DegreeProg is another key.

We would now be able to mention some objective facts about keys, which legitimize the significance ascribed to them. In any case, we can take note of how each connection dependably has a key. A connection is a set comprised of components that are not quite the same as each other. As an outcome, for every connection r(X), the set X of the considerable number of properties is a super-key for it. Presently there are two cases:

(a) a set is additionally a key, so affirming the presence of the key itself, or (b) a key, on the grounds where there exists another super-key contained inside it. At that point, we can continue by rehashing a similar contention on this new set. Since the arrangement of properties engaged with a connection is limited, the procedure ends in a limited number of ventures with a negligible super-key. In this manner, we can presume that each connection has a key.

No less than one key can be characterized for every connection. This ensures access to every one of the values of a database and their unambiguous distinguishing proof.

In addition, it permits the compelling foundation of the connections between information contained in various relations, which portrays the social model as an 'esteem based' model. Let us take a look at the case in Figure 8. In the EXAMS connection, reference is made to the students through the RegNum, while through Code in the COURSES. In actuality, RegNum is the key of the STUDENTS connection and Code is the key of the COURSES connection. So the values of the key credits are in fact used to allude to the substance of each of the relations from the outside (that is, in different relations).

Keys and Invalid Values

We can now come back to our discussion previously regarding the need of staying away from the multiplication of invalid values in our relations. Specifically, we will take note of how, within the sight of invalid values for the key values. It is not true anymore that the values of these keys allow the unambiguous identification of this tuples of the relations and to build up associations between tuples of various relations. Consider the connection in Figure 19, characterized on an indistinguishable pattern from the connection in Figure 18. There are two keys, one of which is made up of the sole characteristic RegNum and the other of the traits Surname, FirstName, and BirthDate. The first tuple has invalid values under RegNum and BirthDate. Further, the table shows that there is no unique value in any of the characteristics. Hence, this tuple is not identifiable in any conceivable route. Specifically, in the event that we need to embed another tuple into the database identifying with a student name John Smith, we cannot know whether we are in reality alluding to a similar student or to another one. Moreover, it is unrealistic to allude to this tuple in different relations of the database, since this must be finished by means of the values of a key. The last two tuples likewise introduce an issue. Taking into consideration that each of them has NULL value (BirthDate and DegereeProg in the third, while RegNum in the fourth), retrieving the records through Surname and FirstName will be still confusing. It is difficult to know whether the two tuples allude to two distinct students or the same.

RegNum	Surname	FirstName	BirthDate	DegreeProg
NULL	Smith	John	NULL	Computing
587614	Smith	Lucy	01/05/61	Engineering
934856	Black	Lucy	NULL	NULL
NULL	Black	Lucy	05/03/58	Engineering

Figure 19 – Invalid Values in All Keys

(Source: Teorey, T. J. Design of Database Structures.)

The case unmistakably recommends the need of restricting the presence of invalid values in the keys of relations. Practically, we receive a straightforward arrangement, which makes it conceivable to ensure the unambiguous distinguishing proof of each tuple and allude to it from within different relations. Invalid values are illegal and not permitted on one of the keys (called the essential key) and for the most part (that is, unless determined otherwise) permitted on the others. The properties that make up the essential key are regularly underlined, as appeared in Figure 20. The greater parts of the references between relations are acknowledged through the estimations of the essential key.

RegNum	Surname	FirstName	BirthDate	DegreeProg
643976	Smith	John	NULL	Computing
587614	Smith	Lucy	01/05/61	Engineering
934856	Black	Lucy	NULL	NULL
735591	Black	Lucy	05/03/58	Engineering

Figure 20 – Primary Key Relation

(Source: Teorey, T. J. Design of Database Structures.)

It is significant that in most genuine cases, it is conceivable to discover qualities whose values are accessible for an essential key. Notwithstanding, at times this does not occur and it winds up noticeably important to present an extra code characteristic that is produced and connected with each tuple at the season of inclusion. Note that many recognizing codes (counting, for instance, student enlistment numbers and government disability numbers) were presented in the past, before the innovation or the spread of the adoption of the database concept, to ensure the unambiguous distinguishing proof of the subject of a domain or area (individually the students and the subjects) and to streamline the reference to them – unequivocally the objectives of keys.

Referential Imperatives and Constrains

Keeping in mind the end goal to talk about the most vital class of between inter-relational requirements, let us consider the database shown in Figure 21. In it, the first connection contains data identifying with an arrangement of traffic-movement offenses, the second one on police officers who have made the charges, while the third on arrangement of engine cars.

The data in the OFFENCES connection is given importance and culmination through some kind of reference to the other two relations. Between OFFENSES and OFFICERS, a relationship is between Officer in OFFENCES and RegNum in OFFICERS. While between OFFENCES and CARS, a relationship is between Department and Registration both in OFFENCES and CARS. The references are important in that the values in the connection OFFENCES are equivalent to values really shown in the other two. If a value of Officer in Offenses does not show up as a value of the specific key of OFFICERS, the reference is most certainly not powerful (thus pointless). In the case we are considering, every one of the references is really usable.

A referential limitation (or remote key) between an arrangement of traits X of a connection R1 and another connection R2 is fulfilled if the qualities in X of each tuple of the example of R1 show up as values-estimations of the (essential) key of the case of R2. The exact definition requires a little care, especially for the situation in which the key of the connection alluded to comprise more than one characteristic, also for the situation in which there is more than one key. We will continue in stages, taking a gander at the case in which the key of R2 is one of a kind and comprises of a sole trait B (and along these lines the set X is actually in its turn made of a sole property A). At this point, the referential imperative between the characteristic An of R1 and the connection R2 is fulfilled if for each tuple t1 in R1 with the end goal that t1[A] is not invalid, there exists a t2 tuple in R2 with the end goal that t1[A] = t2[B]. Usually, we should assess the way that each of the properties in X must relate to an exact trait of the essential key K of R2. It is important to indicate a request both in the set of X and in K. Showing the qualities all together, X = A1A2… Ap and K = B1B2… Bp, the imperative is fulfilled if for each tuple t1 in R1 without any nulls in the X, there exists a t2 tuple in R2 with t1[Ai]=t2[Bi], for $1 \leq i \leq p$.

OFFENCES

Code	Date	Officer	Department	Registration
143256	25/10/92	567	75	5694 FR
987554	26/10/92	456	75	5694 FR
987557	26/10/92	456	75	6544 XY
630876	15/10/92	456	47	6544 XY
539856	12/10/92	567	47	6544 XY

OFFECERS

RegNum	Surname	FirstName
567	Brun	Jean
456	Larue	Henri
638	Larue	Jacques

CARS

Registration	Department	Owner	Address
6544 XY	75	Cordon Edouard	Rue du Pont
7122 HT	75	Cordon Edouard	Rue du Pont
5694 FR	75	Latour Hortense	Avenue Foch
6544 XY	47	Mimault Bernard	Avenue FDR

Figure 21 – Referential Limitations in a Database

(Source: Teorey, T. J. Design of Database Structures.)

On the mapping of the database in Figure 21, it bodes well to characterize the referential imperatives:

- between the Officer quality of the OFFENCES connection and the OFFICERS connection; and

- between the qualities Registration and Department of OFFENCES and the CARS connection, in which the request of the qualities in the key sees first the Registration and afterward Department.

The database in Figure 21 fulfils the two limitations, though the database in Figure 22 violates and disregards both. The infringements or disregards are as follows: the OFFICERS connection does not contain a tuple with the value of 456 in the RegNum; and the CARS actually does not have tuple with 75 in Department with 6544 XY in the Registration as found in the OFFENCES table.

OFFENCES

Code	Date	Officer	Department	Registration
987554	26/10/92	456	75	6544 XY
630876	15/10/92	456	47	6544 XY

OFFECERS

RegNum	Surname	FirstName
567	Brun	Mavis
638	Larue	Jacques

CARS

Registration	Department	Owner	Address
7122 HT	75	Cordon Edouard	Rue du Pont
5694 FR	93	Latour Hortense	Avenue Foch
6544 XY	47	Mimault Bernard	Avenue FDR

Figure 22 – Referential Limitations and a Database that Violates Them

(Source: Maier, D. The Theory of Relational Databases. Potomac, Md.)

With reference to the second limitation, the dialog about the request of the traits can seem over the top, as the correspondence can be accomplished by methods for the names of the properties themselves. Generally, however, this requirement is not conceivable and in this way the requesting is fundamental. Let us illustrate the problem with a database containing data about vehicles engaged in street accidents. Specifically, let us assume that we need to incorporate into a connection, together with other data, the **Registration** part and **Department** part of each of the two vehicles involved.

For this situation, we should have two sets of characteristics, also two referential imperatives. For instance, the outline could be:

ACCIDENTS (Code, Dept1, Registration1, Dept2, Registration2…)

For this situation, it will clearly not be conceivable to build up the correspondence in the referential requirement to the connection CARS by methods for the names of the properties, in that they are unique in relation to those of the essential key CARS. Just by methods for the requesting does it end up noticeably conceivable to determine that the reference partners Dept1 in ACCIDENTS to Department in CARS, Registration1 in ACCIDENTS to Registration in CARS, Dept2 in ACCIDENTS to Department in CARS, and Registration2 in ACCIDENYT to Registration in CARS. The database in Figure 23 fulfils the two requirements, while the one in Figure 24 fulfils the one identified in Dept1 and Registration1.

ACCIDENTS

Code	Dept1	Registration1	Dept2	Registration2	…
6207	75	6544 XY	93	9775 GF	…
6974	93	5694 FR	93	9775 GF	…

CARS

Registration	Department	Owner	Address
7122 HT	75	Cordon Edouard	Rue du Pont
5694 FR	93	Latour Hortense	Avenue Foch
9775 GF	93	LeBlance Pierre	Rue du la Gare
6544 XY	75	Mimault Bernard	Avenue FDR

Figure 23 – Two Different Referential Limitations in a Database

(Source: Maier, D. The Theory of Relational Databases. Potomac, Md.)

ACCIDENTS

Code	Dept1	Registration1	Dept2	Registration2	…
6207	75	6544 XY	93	9775 GF	…
6974	93	5694 FR	93	9775 GF	…

CARS

Registration	Department	Owner	Address
7122 HT	75	Cordon Edouard	Rue du Pont
5694 FR	93	Latour Hortense	Avenue Foch
6544 XY	75	Mimault Bernard	Avenue FDR

Figure 24 - Referential Limitations and a Database that Violates Them

(Source: Maier, D. The Theory of Relational Databases. Potomac, Md.)

A last perception may be valuable in regards to relations with more than one key. For this situation, one of the keys ought to be demonstrated as the essential key, and it is sensible that the references ought to be coordinated towards it. For this reason, in the particular of the referential imperatives, we have been capable to overlook the unequivocal specify of the properties that make up the essential key. Also, it ought to be noticed that not all DBMSs available in the market permit the express sign of the essential key. Some permit the particular of something beyond one key; however, not featuring one as an essential key. In this case, the referential imperative must show expressly the traits that make up the way to which it alludes. For instance, let us consider a database on the mapping:

EMPLOYEES (EmpNum, Surname, FirstName, Department)

DEPARTMENTS (Code, Name. Location)

in which the DEPARTMENTS connection is distinguished by the Code characteristic and independently by the Name quality (no two offices exist with the same code or with a similar name). It is helpful that one of the two keys, for-example Code, is shown as an essential, and used to set up references. In the event that, nonetheless, the framework does not take into account the idea of an essential key, the limitation must be communicated by the unequivocal sign of the characteristics. We should in this manner say that there exists a referential imperative between the Department characteristic of the EMPLOYEES connection and the Code attribute of the DEPARTMENTS connection.

CONCLUSIONS

In this part we have characterized the structures and limitations of the relational model. Initially we talked about the idea of connection, with a few varieties concerning the ideas of set hypothesis. Then, we demonstrated how relations can be utilized to sort out more mind boggling accumulations of information utilizing the information itself to make references between various parts (without the utilization of express pointers). At that point, in the light of presenting the need for utilizing invalid qualities to indicate the nonattendance of data, we examined the idea of uprightness limitations through three essential classes: tuple imperatives, keys, and referential requirements(Atzeni, 1993).

In the following two sections, we finish the introduction of the relational model from two perspectives.

- In Chapter 3, we delineate the establishments of question dialects, that is, the languages used to get to data in databases.
- While, in Chapter 4, we indicate how every one of the ideas, those with respect to the structures and to the requirements, as talked about in this part, and those with respect to query languages (Chapter 3) are executed in business DBMSs, utilizing SQL.

CHAPTER 3

ALGEBRA AND CALCULUS OF RELATIONAL DATABASE

We have found in the past parts of the book that data of importance to information administration applications can be delivered by methods for relations. The languages for indicating operations for questioning/ querying and refreshing the information itself constitute, in their turn, a basic segment of every information display. A refresh can be viewed as a capacity that, given a database, creates another database (without changing the whole schema). Then, a query can additionally be considered as a capacity that, given a database, delivers a connection.

Along these lines, either to cross-examine or to refresh the database, we have to build up the capacity to express types of functions differently on the database. It is critical to understand the establishments and fundamentals of queries and refresh languages and after that apply those establishments when considering the languages that are really upheld by business DBMSs.

We will see the relational algebra. This is actually a procedural language (that is, one in which the information recovery capacities are determined by portraying the technique that must be followed to get the outcome). We will delineate the different administrators of the variable based algebra. The way operators can be consolidated to frame articulations and the methods by which articulations can be changed to enhance productivity. We will likewise portray the impact that null values have on the social variable based algebra, and afterward how a question language can be utilized to characterize virtual relations (otherwise called sees or views), which are most certainly not to put away in the database.

In connection, we will give a compact introduction of relational algebra, an explanatory language, in which the information recovery capacities portray the properties of the outcome, as opposed to the method used to acquire it. This language depends on the first request predicate analytics and we will show two forms: the primary specifically got from predicate analytics and the second that endeavours to conquer a portion of the restrictions of the first.

We will finish up the part with a short treatment of Datalog, an intriguing commitment from late research, which permits the detailing of queries that could not be communicated in polynomial algebra or in analytics. The segments on analytics and Datalog can be excluded without trading off the comprehension of the succeeding parts.

In the following section, committed to SQL, we will perceive how it can be helpful to consolidate explanatory and procedural angles inside a solitary language. We will likewise perceive how refreshes depend on indistinguishable standards from queries (Codd E. , 1971).

RELATIONAL ALGEBRA

As we have said, relational algebra is a procedural type of language, in view of arithmetical ideas. It comprises of a set of operations that are characterized on relations, and that create relations as results. Along this, we can build articulations that include more than one operation, keeping in mind the end goal to plan complex type of queries. In the accompanying segments, we inspect the different operations.

- To start with, those of customary set hypothesis, union, crossing point, distinction.
- Next, the more particular ones, renaming, determination, projection.
- Lastly, the most imperative, the join, in its different structures, characteristic join, Cartesian item and theta-join.

Union, Crossing Point, Distinction

In any case, take note that relations are sets. So it bodes well to characterize for them, the conventional set of operations such as union, distinction, and crossing point.

We should also know about the way that a connection is not non-exclusively a set of tuples, yet it is an arrangement of homogeneous tuples, which are, tuples characterized on the same traits. So, regardless of the possibility that it could be done on a fundamental level, to characterize these operations on any combination of relations, there is no sense from the perspective of the relational model in characterizing them with reference to relations on various characteristics/attributes. For instance, the union of two relations r_1 and r_2 on various patterns would be an arrangement of heterogeneous tuples; some characterized on the characteristics of r_1 and the others are on those of r_2. This would actually be inadmissible since an arrangement of heterogeneous tuples is not a connection, so as to consolidate the operations to form complex articulations we need the outcomes to be relations. Along these lines, in relational algebra, we permit the use of operations, namely union, crossing point, and distinction to sets of relations characterized on similar traits. Figure 25 demonstrates the cases on the use of the three operations, with the standard definitions, adjusted to our specific situation.

- The union of two relations r_1 (X) and r_2 (X), characterized on a similar arrangement of qualities X, is communicated as $r_1 \cup r_2$ and is additionally a connection on X containing the tuples that have a place with r_1 or to r_2, or to both.
- The distinction (difference) of r_1 (X) and r_2 (X) is communicated as $r_1 - r_2$ and is a connection on X containing the tuples that have a place with r_1 and not with r_2.
- The crossing point (intersection) of r_1 (X) and r_2 (X) is communicated as $r_1 \cap r_2$ and is a connection on X containing the tuples that have a place with both r_1 and r_2.

GRADUATES

Number	Surname	Age
7274	Robinson	37
7432	O'Malley	39
9824	Darkess	38

MANAGERS

Number	Surname	Age
9297	O'Malley	56
7432	O'Malley	39
9824	Darkess	38

GRADUATES ∪ MANAGERS

Number	Surname	Age
7274	Robinson	37
7432	O'Malley	39
9824	Darkess	38
9297	O'Malley	56

GRADUATES ∩ MANAGERS

Number	Surname	Age
7432	O'Malley	39
9824	Darkess	38

GRADUATES MANAGERS

Number	Surname	Age
7274	Robinson	37

Figure 25 - Union, Difference, and intersection Examples.

(Source: Maier, D. The Theory of Relational Databases.)

Renaming

The restrictions we have needed to force on the standard set of operations, even though justifiable, appear to be especially restrictive. For example, consider the two relations in Figure 26. It is significant to execute a kind of union on them with a specific end goal to get all the "parent–child" sets held in the database. In any case, that is unrealistic, in light of the fact that the trait that we have intuitively called Parent, is in certainty called Father in one connection and Mother in the other.

PATERNITY

Father	Child
Adam	Cain
Adam	Abel
Abraham	Isaac
Abraham	Ishmael

MATERNITY

Mother	Child
Eve	Cain
Eve	Seth
Sarah	Isaac
Hagar	Ishmael

PATERNITY ∪ MATERNITY ??

Figure 26 – An Incorrect Union though Meaningful.

(Source: Maier, D. The Theory of Relational Databases.)

To determine the issue, we present a particular operation, in which the sole design is to adjust property names, as fundamental, to encourage the application certain operations. The operation is called renaming; on the grounds that it really changes the names of the characteristics, leaving the substance of the relations unaltered. A case of renaming is appeared in Figure 27. The operation changes the name of the credit Father to Parent, as demonstrated by the documentation Parent ← Father and given in subscript of Q symbol, which means the renaming. Taking a gander at the table, it is anything but difficult to perceive how just the heading changes, clearing out the principle body unaltered.

PATERNITY

Father	Child
Adam	Cain
Adam	Abel
Abraham	Isaac
Isaac	Jacob

$Q_{\text{Parent} \leftarrow \text{Father}}$(PATERNITY)

Parent	Child
Adam	Cain
Adam	Abel
Abraham	Isaac
Isaac	Jacob

Figure 27 – Renaming.

(Source: Maier, D. The Theory of Relational Databases.)

Figure 28 demonstrates the use of the union to the consequence of two renamings of the relations in Figure 26.

$Q_{\text{Parent} \leftarrow \text{Father}}$(PATERNITY) $\cup$ $Q_{\text{Parent} \leftarrow \text{Mother}}$(MATERNITY)

Parent	Child
Adam	Cain
Adam	Abel
Abraham	Isaac
Abraham	Ishmael
Eve	Cain
Eve	Seth
Sarah	Isaac
Hagar	Ishmael

Figure 28 – Union with Two Renaming.

(Source: Maier, D. The Theory of Relational Databases.)

Let us characterize the renaming operation in general terms. Give r a chance to be a connection characterized on the arrangement of characteristics X and let Y be another arrangement of traits with a similar cardinality. Let A1A2… Ak and B1B2… Bk be separately a request of the characteristics in X and a request of those in Y.

At that point the renaming

$$Q_{B_1 B_2 \ldots B_k \longleftarrow A_1 A_2 \cdots A_k}(r)$$

contains a tuple t' for each tuple t in r, characterized as takes after. t' is a tuple on Y and t'[Bi]=t[Ai] for i = 1, … , n. The definition affirms that the progressions that happen are changes to the names of the traits, while the qualities stay unaltered and are related with the new characteristics. Practically, in the two records $A_1A_2\ldots A_k$ and $B_1B_2\ldots B_k$ we show how those traits are renamed (that is, for those which Ai ≠ Bi). This is the motivation behind why in Figure 27 we have composed

$$Q_{Parent \leftarrow Father}(PATERNITY)$$

and not

$$Q_{Parent, Child \leftarrow Father, Child}(PATERNITY)$$

Figure 29 demonstrates another case of union which was applied after renaming. In this case, in every connection there are two traits that are renamed and subsequently the request of the sets (Branch, Salary et cetera) is critical.

EMPLOYEES

Surname	Branch	Salary
Patterson	Rome	45
Trumble	London	53

STAFF

Surname	Factory	Wages
Cooke	Chicago	33
Bush	Monza	32

$Q_{Location, Pay \leftarrow Branch, Salary}(EMPLOYEES) \cup Q_{Location, Pay \leftarrow Factory, Wages}(STAFF)$

Surname	Location	Pay
Patterson	Rome	45
Trumble	London	53
Cooke	Chicago	33
Bush	Monza	32

Figure 29 – Renaming and Union.

(Source: Ceri, S. Methodology and Tools for Database Design)

Selection

We now turn our consideration regarding the operations of relational algebra that permit the control of relations. There are three operations: selection, projection, and join (the last having a few variations).

Before expounding, take note that selection and projection enable capacities that could be characterized as complementary (or differently called orthogonal). They are actually both unary (which means, they have one connection as argument) and create as result a bit/part of that connection. Specifically, a choice creates a subset of tuples on every one of the properties, while a projection gives an outcome to which all the tuples contribute, yet on a subset of qualities. As presented in Figure 30, we can state that selection produces 'horizontal disintegrations' and projection produces 'vertical disintegrations' (Paredaens, The Structure of the Relational Database, 1990).

Figure 31 and Figure 32 demonstrate two cases of selection that outline the central qualities of the operation, meant by the image σ, with the fitting 'selection condition' showed as subscript. The outcome contains the tuples of the operand that fulfil the condition. As appeared in the illustrations, the choice conditions can permit both for correlations among traits and for examinations among properties and constants, and can be complex, being gotten by joining basic conditions with the coherent connectives ∨ (or), ∧ (and) and ¬ (not).

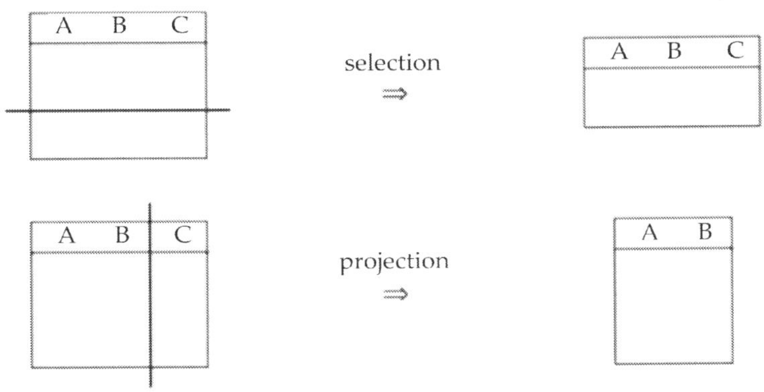

EMPLOYEES

Surname	FirstName	Age	Salary
Smith	Mary	25	2000
Black	Lucy	40	3000
Verdi	Nico	36	4500
Smith	Mark	40	3900

$\sigma_{Age<30 \vee Salary>4000}$(EMPLOYEES)

Surname	FirstName	Age	Salary
Smith	Mary	25	2000
Verdi	Nico	36	4500

Figure 31 – Selection.

(Source: Ceri, S. Methodology and Tools for Database Design)

Given a connection, a propositional equation F on X is a recipe acquired by joining nuclear states of the sort AϑB or even Aϑc with the connectives which are '$\vee$', '$\wedge$', and '$\neg$', where:

- ϑ is actually a correlation operation ($=, \neq, >, <, \geq, \leq$),
- A and also B are characteristics in X that are perfect (that is, the correlation ϑ is significant on the values of their domains), and
- c is a constant perfectly compatible in the domain of A.

Given a recipe F and a tuple t, a true value is characterized for F on t.

EMPLOYEES

Surname	FirstName	Age	Salary
Smith	Mary	25	2000
Black	Lucy	40	3000
Verdi	Nico	36	4500
Smith	Mark	40	3900

$\sigma_{Age<30 \vee Salary>4000}$(EMPLOYEES)

Surname	FirstName	Age	Salary
Smith	Mary	25	2000
Verdi	Nico	36	4500

Figure 32 – Selection (Another).

(Source: Ceri, S. Methodology and Tools for Database Design)

- $A\vartheta B$ is valid on t if and just if t[A] is in connection ϑ with t[B] (for case, A=B is valid on t if and just if t[A]=t[B]).
- $A\vartheta c$ is valid on t if and just if t[A] is in connection ϑ with the c.
- $F_1 \vee F_2$, also $F_1 \wedge F_2$ and $\neg F_1$ have the typical meaning.

Now we can finish the definition.

- The selection σF(r) produces a connection on indistinguishable traits from r that contains actually the tuples of r for which the F is valid.

Projection

The meaning of the projection operation is quite simple. Given a connection r(X) and a subset Y of X, the projection of r on Y (demonstrated by πY(r)) is the arrangement of tuples on Y got from the tuples of r considering just the qualities on Y.

$$\pi_Y(r) = \{t[Y] \mid t \in r\}$$

Figure 33 demonstrates the first case of projection that unmistakably represents the idea specified previously. The projection permits the vertical disintegration of relations. The consequence of the projection contains for this situation the same number of tuples as its operand, characterized however just on a portion of the qualities.

EMPLOYEES

Surname	FirstName	Department	Head
Smith	Mary	Sales	De Rossi
Black	Lucy	Sales	De Rossi
Verdi	Mary	Personnel	Fox
Smith	Mark	Personnel	Fox

$\pi_{Surname,FirstName}$(EMPLOYEES)

Surname	FirstName
Smith	Mary
Black	Lucy
Verdi	Mary
Smith	Mary

Figure 33 - Projection.

(Source: Ceri, S. Methodology and Tools for Database Design)

Figure 34 demonstrates another projection, in which we take note of an alternate circumstance. The outcome contains less tuples than the operand, since all the actual tuples in the operand that do have break-even with values on every one of the properties of the projection give a similar commitment to the projection itself. As relations are characterized as sets, they are not permitted to have tuples with similar values: rise to commitments "fall" into a solitary tuple.

EMPLOYEES

Surname	FirstName	Department	Head
Smith	Mary	Sales	De Rossi
Black	Lucy	Sales	De Rossi
Verdi	Mary	Personnel	Fox
Smith	Mark	Personnel	Fox

$\pi_{\text{Department, Head}}$(EMPLOYEES)

Department	Head
Sales	De Rossi
Personnel	Fox

Figure 34 – Projection with More Operands than Tuples

(Source: Silberschatz. Database System Concepts.)

We can state that the aftereffect of a projection contains at most as numerous tuples as the operand, but can contain less as appeared in Figure 34. Note additionally that there exists a connection between the key requirements and the projections. $\pi Y(r)$ contains an indistinguishable number of tuples from r if and just if Y is a super-key for r. So:

- if Y is a super-key, at that point r does not contain sets of tuples that are equivalent on Y, and in this manner each tuple makes an alternate commitment to the actual projection; and
- if the projection does have the same number of tuples as the operand, at that point each tuple of r adds to the projection with various values, and accordingly r does not contain sets of tuples break even with on Y; however, this is precisely the meaning of a super-key.

For the EMPLOYEES connection in Figure 33 and Figure 34, the qualities Surname and FirstName shape a key (and along these lines a super-key), while Department and Head do not frame a super-key. Take note that a projection can create various tuples equivalent to those of the operand regardless of the possibility that the qualities included are not characterized as super-keys (of the outline) but happen to be a super-key for the particular connection. For instance, on the off chance that we rethink the relations talked about in Chapter 2 on the diagram

STUDENTS(RegNum, Surname, FirstName, BirthDate, DegreeProg)

we can state that for every one of the relations, the real projection on RegNum and that on Surname, FirstName, and BirthDate have an indistinguishable number of tuples from the operand. On the other hand, a projection on Surname and DegreeProg can have less tuples. However, in the specific case in which there are no students with a similar surname enlisted on the same degree program, the projection on Surname and DegreeProg additionally has an indistinguishable number of tuples from the operand.

Join

Let us analyse the join operation that is the most imperative one in relational algebra. The join enables us to set up associations among information contained in various relations, contrasting the qualities contained in them and utilizing the central qualities of the model. There are two fundamental variants of the operations, which are in any case, realistic one from the other. The first is valuable for a presentation and the second is maybe more significant from a pragmatic perspective (Maier D. , 1984).

Natural Join

The natural join, indicated by the image $\bowtie$, is an operation that corresponds information in various relations, on the premise of equivalent values of traits with a similar name. (The join is characterized here with two operands, in any case, can be summed up.) Figure 35 demonstrates a case on this. The consequence of the join is a connection on the union of the arrangements of properties of the operands. Into the fig, the outcome is characterized on Employee, Department, and Head, that is, on the union of the Employee and Department, and of the Department and Head. The tuples in the join can be taken by joining the tuples of the operands with measure up to esteems-values on the regular qualities. In the illustration involving the characteristic Department, the main tuple of the join is gotten from the blend of the first tuple of the connection r_1 and the second tuple of r_2. In certainty, both have deals as the incentive for Departments.

r_1

Employee	Department
Smith	sales
Black	production
Bianchi	production

r_2

Department	Head
production	Mori
sales	Brown

$r_1 \bowtie r_2$

Employee	Department	Head
Smith	sales	Brown
Black	production	Mori
Bianchi	production	Mori

Figure 35 – Regular Join (Natural).

(Source: Silberschatz, A. K. *Database System Concepts.* New York)

As a rule, we say that the natural join $r_1 \bowtie r_2$ of $r_1(X_1)$ and r_2 (X_2) is a connection characterized on X_1, X_2 (which is, on the actual union of the sets of X_1 and X_2), followed by:

$$r_1 \rhd\lhd r_2 = \{t\ on\ X_1X_2 \mid exist\ t_1 \in r_1\ and\ t_2 \in r_2 with\ \ t[X_1] = t_1\ and\ t[X_2]$$

And more significantly, we could have composed:

$$r_1 \rhd\lhd r_2 = \{t\ on\ X_1X_2 \mid t_1[X_1] \in r_1\ and\ t_2[X_2] \in r_2\}$$

The definition affirms that the tuples of the outcome are gotten by joining tuples of the operands with break even with values on the regular characteristics. On the off chance that we demonstrate the basic traits as $X_1;2$ ($X_{1;2} = X_1 \cap X_2$), at that point the two conditions which are $t[X_1]=t_1$ and $t[X_2]=t_2$ suggest (since $X_{1,2} \subseteq X_1$ also, $X_{1,2} \subseteq X_2$) that $t[X_1,2]=t_1[X_{1,2}]$ and $t[X_{1,2}] = t_2[X_{1,2}]$ and along these lines $t_1[X_{1,2}] = t_2[X_{1,2}]$. The level of the aftereffect of a join is not exactly or equivalent to the total of these degrees of the two operands, on the grounds that the basic traits of the operands seem just once in the outcome.

Note that regularly, the normal traits in a join shape the key of 1 of the relations. In most cases, there is additionally a referential limitation between the basic traits. Note that each of the tuples in OFFENCES has been joined with precisely 1 of the tuples of CARS: (i) at most 1 since

Department and Registration shape a key for CARS; (ii) no less than one on account of the referential imperative between Department in CARS and Registration in OFFENCES. The join, in this manner, has precisely the same number of tuples as the connection OFFENCES (Atzeni, 1993).

OFFENCES

Code	Date	Officer	Department	Registration
143256	25/10/92	567	75	5694 FR
987554	26/10/92	456	75	5694 FR
987557	26/10/92	456	75	6544 XY
630876	15/10/92	456	47	6544 XY
539856	12/10/92	567	47	6544 XY

CARS

Registration	Department	Owner	Address
6544 XY	75	Cordon Edouard	Rue du Pont
7122 HT	75	Cordon Edouard	Rue du Pont
5694 FR	75	Latour Hortense	Avenue Foch
6544 XY	47	Mimault Bernard	Avenue FDR

OFFENCES ⋈ CARS

Code	Date	Officer	Depart-ment	Regist-ration	Owner	Address
143256	25/10/92	567	75	5694 FR	Latour Hortense	Avenue Foch
987554	26/10/92	456	75	5694 FR	Latour Hortense	Avenue Foch
987557	26/10/92	456	75	6544 XY	Cordon Edouard	Rue du Pont
630876	15/10/92	456	47	6544 XY	Mimault Bernard	Avenue FDR
539856	12/10/92	567		6544 XY	Mimault Bernard	Avenue FDR

Figure 36 - The Relations between the OFFENCES and the CAR.

(Source: Silberschatz, A. K. Database System Concepts. New York)

Figure 37 demonstrates another case of join, utilizing an indistinguishable relation that we have effectively utilized to show a union which was preceded actually by renaming. Here, the information of the two relations is joined by the estimation (value) of the child, restoring the guardians for every individual shown in the database.

PATERNITY

Father	Child
Adam	Cain
Adam	Abel
Abraham	Isaac
Abraham	Ishmael

MATERNITY

Mother	Child
Eve	Cain
Eve	Seth
Sarah	Isaac
Hagar	Ishmael

PATERNITY ⋈ MATERNITY

Father	Child	Mother
Adam	Cain	Eve
Abraham	Isaac	Sarah
Abraham	Ishmael	Hagar

Figure 37 – Child with both Parents.

(Source: Silberschatz, A. K. Database System Concepts. New York)

The two illustrations show how the different relational algebra operations permit distinctive methods for joining and associating the information contained in a database, as indicated by the different necessities.

Semijoin

Let us look at some extraordinary illustrations of joins keeping in mind the end goal to feature and highlight some imperative points and areas of focus. For the case shown in Figure 35, we can state that each tuple of each of the operands does contribute to no less than one tuple of the outcome. For this situation, the join is said to be finished.

For each t1 tuple of r_1, there is a t tuple in $r_1 \bowtie r_2$ with the end goal that $t[X_1] = t_1$ (and thus for r_2). This property does not hold by and large, since it actually requires a correspondence and correlation between the tuples of the two relations. Figure 38 demonstrates a join where some tuples that are into operands (specifically, the 1st of r_1 and the 2nd of r_2) do not add to the outcome. This is on the account of these tuples having no partner (that is, a tuple with a similar incentive on the normal quality Department) in the other connection. These tuples are also called as *dangling tuples*.

r_1

Employee	Department
Smith	sales
Black	production
White	production

r_2

Department	Head
production	Mori
purchasing	Brown

$r_1 \bowtie r_2$

Employee	Department	Head
Black	production	Mori
White	production	Mori

r_1

Employee	Department
Smith	sales
Black	production
White	production

r_2

Department	Head
marketing	Mori
purchasing	Brown

$r_1 \bowtie r_2$

Employee	Department	Head

Figure 39 – Empty Join.

(Source: Silberschatz, A. K, Database System Concepts)

In the outrageous inverse circumstance, each tuple of every operand can be consolidated with all the tuples of the alternate. In this case, the outcome contains various tuples equivalent to the result of the cardinalities of the operands and in this manner, $|r_1| \times |r_2|$ tuples (where $|r|$ demonstrates the cardinality of the connection r).

r_1

Employee	Project
Smith	A
Black	A
White	A

r_2

Project	Head
A	Mori
A	Brown

$r_1 \bowtie r_2$

Employee	Project	Head
Smith	A	Mori
Black	A	Mori
White	A	Mori
Smith	A	Brown
Black	A	Brown
White	A	Brown

Figure 40 – Join $|r_1| \times |r_2|$.

(Source: ElMasri, R. A. Fundamentals of Database Systems.)

To condense, we can state that the join of r_1 and r_2 contains various tuples in the vicinity of zero and $| r_1| \times | r_2|$. Moreover,

- in case the join of r_1 and also r_2 is finished, at that point it contains various tuples at slightest equivalent to the greatest of $| r_1|$ and also $| r_2|$;
- a case that $X_1 \cap X_2$ contains a key for r_2, at that point the join of the r_1 (X_1) and the r_2 (X_2) contains at maximum $| r_1|$ tuples; and
- if $X_1 \cap X_2$ is the essential key for the r_2 and there is a referential requirement between $X_1 \cap X_2$ in r_1 and such a key of the r_2, at that point the join of the r_1 (X_1) and the r_2 (X_2) contains precisely $| r_1|$ tuples.

Outer Joins

The way that the join operation 'forgets' the tuples of a connection that have no partner in the other operand is valuable sometimes in different cases, but not so valuable in other cases given the likelihood of excluding imperative and important data. For instance, let us look at the join in Figure 41. Let us assume that we are intrigued by every one of the workers, alongside their separate heads if known.

The characteristic join would not help in creating this outcome. For this reason, a variation of the operation called external join was proposed (and embraced in the last adaptation of the SQL). This takes into account the likelihood that all the tuples add to the outcome, reaching out with invalid esteems (values) where there is no partner. There are three variations of this operation: the left external join that broadens just the tuples of the main operand, the right external join that broadens those of the second operand, and the full external join that expands all tuples. In Figure 41 we exhibit cases of external joins on the relations as found in Figure 38. The syntax is as clear as crystal.

r_1

Employee	Department
Smith	sales
Black	production
White	production

r_2

Department	Head
production	Mori
purchasing	Brown

$r_1 \bowtie_{LEFT} r_2$

Employee	Department	Head
Smith	sales	NULL
Black	production	Mori
White	production	Mori

$r_1 \bowtie_{RIGHT} r_2$

Employee	Department	Head
Black	production	Mori
White	production	Mori
NULL	purchasing	Brown

$r_1 \bowtie_{FULL} r_2$

Employee	Department	Head
Smith	sales	NULL
Black	production	Mori
White	production	Mori
NULL	purchasing	Brown

Figure 41 – Outer Joints.

(Source: ElMasri, R. A. Fundamentals of Database Systems)

N-ary Join, Crossing Point, and Cartesian Item

Let us take a look at some of the properties of the natural join. (We allude here to natural join instead of to outer join, for which a portion of the properties talked about here do not hold.)

To begin with, let us watch that it is commutative, that is, $r_1 \bowtie r_2$ is constantly equivalent to $r_2 \bowtie r_1$, and acquainted, $r_1 \bowtie (r_2 \bowtie r_3)$ is equivalent to $(r_1 \bowtie r_2) \bowtie r_3$. Hence, we can compose, where important, join successions without sections:

$r_1 \bowtie r_2 \ldots \bowtie r_n$ or $\bowtie^n_{i=1} r_i$

Note likewise that we have expressed no particular theory about the arrangements of qualities X_1 and X_2 on which the operands are characterized. Hence, the two sets could even be equivalent or be disjoint. Give us a chance to analyse these outrageous cases. The general definition given above is as important, yet certain focuses ought to be noted. On the off chance that $X_1 = X_2$, the join concurs with the convergence

$$r_1 (X_1) \bowtie r_2(X_2) = r_1 (X_1) \text{È} r_2(X_2)$$

since, by definition, the outcome is a connection on the union of the two arrangements of traits and must contain the tuples t to such an extent that $t \in r_1$ and also $t \in r_2$, and along these lines matches with the meaning of convergence.

The situation where the two arrangements of properties are disjoint requires significantly more consideration. The outcome is constantly characterized on the union of X_1 and X_2, and each tuple is continuously gotten from two tuples, which are assigned one for each of these operands. In any case, since such tuples have no properties in like manner, there is no necessity to be fulfilled with the goal for them to partake in the join. As previously mentioned, the condition that the tuples must actually have similar esteems (values) on the basic properties is dependably checked. So the aftereffect of this join contains the tuples acquired by consolidating the tuples of the operands in all conceivable ways. For this situation, we regularly say that the join turns into a Cartesian item. This could be depicted as an operation characterized (utilizing a similar definition given above for common join) on relations that have no properties in common. The utilization of the term is marginally deceptive, as it is not by any means the same as a Cartesian item between sets. The Cartesian result of two sets is an arrangement of sets (with the principal component from the primary set furthermore the second). For the situation here we have tuples, each acquired by comparing a tuple of the principal connection and a tuple of the second (Abiteboul, 1995).

Figure 42 demonstrates a case of the Cartesian-item showing how the outcome contains various tuples equivalent to the result of the cardinalities of the actual operands.

Theta-Join and Equi-Join

If we look at Figure 42, clearly a Cartesian item is a rule of almost no utilization, since it consolidates tuples in a way that is not really critical. Truthfully, in any case, the Cartesian item is regularly trailed by a choice that saves just the consolidated tuples that fulfil the given necessities. For instance, it bodes well to characterize a Cartesian item on the relations EMPLOYEES and PROJECTS, in the event that it is taken after by the choice that holds just the tuples with measure up to esteems (values) on the characteristics Project and Code (see Figure 43).

EMPLOYEES

Employee	Project
Smith	A
Black	A
Black	B

PROJECTS

Code	Name
A	Venus
B	Mars

EMPLOYEES ⋈ PROJECTS

Employee	Project	Code	Name
Smith	A	A	Venus
Black	A	A	Venus
Black	B	A	Venus
Smith	A	B	Mars
Black	A	B	Mars
Black	B	B	Mars

Figure 42 – Cartesian Product.

(Source: ElMasri, R. A. Fundamentals of Database Systems.)

Therefore, another operation is frequently presented, the theta-join. It is an inferred operation, as it is characterized by methods for different operations. For sure, it is a Cartesian item took after by a determination, as:

$$r_1 \bowtie_F r_2 = s_F(r_1 \bowtie r_2)$$

EMPLOYEES

Employee	Project
Smith	A
Black	A
Black	B

PROJECTS

Code	Name
A	Venus
B	Mars

$\sigma_{Project=Code}$(EMPLOYEES ⋈ PROJECTS)

Employee	Project	Code	Name
Smith	A	A	Venus
Black	A	A	Venus
Black	B	B	Mars

Figure 43 - Selection Preceded by a Cartesian Product.

(Source: ElMasri, R. A. Fundamentals of Database Systems)

The connection in Figure 43 would thus be able to be gotten utilizing the theta-join:

$$\text{EMPLOYEES} \bowtie_{\text{Project=Code}} \text{PROJECTS}$$

A theta-join in which the state of determination F is a conjunction of iotas of uniformity; each with a trait of the primary connection and one of the second is called equi-join. The connection in Figure 43 was acquired by methods for an equi-join.

From the pragmatic perspective, the theta-join and the equi-join are exceptionally vital. This is on the grounds that most current database frameworks do not take favourable position of ascribe names with a specific end goal to join relations, and in this way utilize the equi-join and theta-join as opposed to the regular join. We inspect this idea more completely when we talk about SQL queries in Chapter 4. The SQL queries for the most part compare to equi-joins, while the common join was made accessible just in the latest adaptations of SQL.

In the meantime, we displayed the normal join first since it permits the straightforward exchange of vital issues, which would then be able to be stretched out to the equi-join.

Note additionally that the normal join can be re-enacted utilizing renaming, equi-join, and projection. Without broadly expounding, here is an illustration.

Given two relations, r_1 (ABC) and r_2 (BCD), the common join of r_1 and r_2 can be communicated by methods for different operations in three stages:

- renaming the credits in order to acquire relations on disjoint diagrams: $Q_{B'C' \leftarrow BC}(r_2)$;
- equi-joining such relations, with balance conditions on the renamed traits: $r_1 \rhd\lhd_{B=B' \wedge C=C'} \left(Q_{B'C' \leftarrow BC}(r_2)\right)$; and
- finishing up with a projection that disposes of all the "copy" properties (one for each match associated with the equi-join): $\pi_{ABCD}\left(r_1 \rhd\lhd_{B=B' \wedge C=C'} \left(Q_{B'C' \leftarrow BC}(r_2)\right)\right)$

Queries in Algebra

Generally, a query can be characterized as a capacity that, when connected to database examples, produces relations. More accurately, given an outline R of a database, a query is a capacity that, for each occurrence r of R, delivers a connection on a given arrangement of characteristics X. The articulations in the different query languages, (for example, relational algebra) "speak to" or "actualize" queries: every articulation characterizes a capacity. We show by methods for E(r) the outcome of the use of the articulation E to the r database.

In relational algebra, queries on a database diagram R are detailed by methods for articulations in which molecules are (the names of) the relations in R (the 'factors'). We finish up the introduction of relational algebra by the definition of few inquiries of expanding many-sided quality, which allude to the outline containing the two relations:

EMPLOYEES(<u>Number</u>, Name, Age, Salary)

SUPERVISION(Head, <u>Employee</u>)

EMPLOYEES

Number	Name	Age	Salary
101	Mary Smith	34	40
103	Mary Bianchi	23	35
104	Luigi Neri	38	61
105	Nico Bini	44	38
210	Marco Celii	49	60
231	Siro Bisi	50	60
252	Nico Bini	44	70
301	Steve Smith	34	70
375	Mary Smith	50	65

SUPERVISION

Head	Employee
210	101
210	103
210	104
231	105
301	210
301	231
375	252

Figure 44 – Database with Expressions.

(Source: Paredaens, J. D. The Structure of the Relational Database.)

A database on such an outline is appeared in Figure 44. The principal query is extremely straightforward, including a solitary connection: discover the numbers, names, and periods of workers gaining more than 40 thousand. For this situation, utilizing a determination, we can feature just the tuples that fulfil the condition and by methods for a projection, kill the undesirable qualities.

$$p_{\text{Number, name, age}}(s_{\text{Salary>40}}(\text{EMPLOYEES})) \quad (3.1)$$

Number	Name	Age
104	Luigi Neri	38
210	Marco Celli	49
231	Siro Bisi	50
252	Nico Bini	44
301	Steve Smith	34
375	Mary Smith	50

The consequence of this articulation, connected to the database in Figure 44, is appeared in Figure 45. The second query includes both the relations in an exceptionally common manner: find the enrolment quantities of the managers of the workers gaining more than 40 thousand.

$$p_{\text{Head}}(\text{SUPERVISION} \bowtie \text{Employee=Number}) (s_{\text{Salary>40}}(\text{EMPLOYEES})) \quad (3.2)$$

The outcome is appeared in Figure 46, alluding again to the database in Figure 44.

Let us proceed onward to some more mind boggling illustrations. We start by marginally changing the above query: discover the names and pay rates of the administrators of the representatives procuring more than 40 thousand. Here, we can clearly utilize the going before articulation, yet we should then deliver, for each tuple of the outcome, the data asked for on the supervisor, which must be separated from the connection EMPLOYEES. Each tuple of the outcome is built on the premise of three tuples, the first from EMPLOYEES (around a worker procuring more than 40 thousand), the second from SUPERVISION (giving the quantity of the chief of the worker being referred to), and the third from the EMPLOYEES (with the data concerning the manager) again. The arrangement naturally requires the join of the connection EMPLOYEES with the aftereffect of the previous articulation, yet a notice is required. Generally speaking, the supervisor and the worker-employee are not the same, and along these lines the two tuples of EMPLOYEES that add to a tuple of the join are extraordinary. The join should along these lines preceded by a reasonable renaming. The accompanying is a case:

$$p_{NameH,SalaryH} \bowtie (Q_{NumberH,NameH,SalaryH,AgeH\ Number,Name,Salary,Age}(EMPLOYEES)$$
$$NumberH{=}Head$$
$$(SUPERVISION \bowtie_{Employee=Number} (s_{Salary>40}(EMPLOYEES))))$$
(3. 3)

Head
210
301
375

Figure 46 –The Results of Expression 3.2.

(Source: Paredaens, J. D, The Structure of the Relational Database.)

The outcome is appeared in Figure 47, again alluding to the database in Figure 44.

NameH	SalaryH
Marco Celli	60
Steve Smith	70
Mary Smith	65

Figure 47 - The Results of Expression 3.3.

(Source: Paredaens, J. D. The Structure of the Relational Database)

The following inquiry is a minor departure from the one above, asking for the correlation of two values of a similar quality, yet from various tuples: discover the representatives procuring more than their particular head, indicating number, name, and salary of the employees and heads. The articulation is like the one above, and the requirement for renaming is likewise obvious. (The result is demonstrated in Figure 48.)

$$p_{Number,Name,Salary,NumberH,NameH,SalaryH}$$

$$(s_{Salary>SalaryH}(Q_{NumberH,NameH,SalaryH,AgeH\ Number,Name,Salary,Age}(EMPLOYEES)$$

$$\bullet\ _{NumberH=Head}(SUPERVISION \bowtie_{Employee=Number} (EMPLOYEES)))) \quad (3.4)$$

Number	Name	Salary	NumberH	NameH	SalaryH
104	Luigi Neri	61	210	Marco Celli	60
252	Nico Bini	70	375	Mary Smith	65

Figure 48 - The Results of Expression 3.4.

(Source: Paredaens, J. D. The Structure of the Relational Database.)

The last illustration requires significantly more care: discover the number and name of the heads whose employees earn more than 40-thousand. The query incorporates some kind of widespread evaluation, yet relational algebra does not contain any build straightforwardly suited to this case. We can continue with a twofold refutation, finding the heads none of whose employees earn 40-thousand or less. This query is conceivable in relational algebra, utilizing the distinction operation.

We select every one of the heads aside from the individuals who have an employee who earns 40-thousand or less. The articulation is as per the following:

$p_{Number,Name}(EMPLOYEES) \bowtie_{NumberH=Head}$

$p_{Head}(SUPERVISION)$ -

$p_{Head}(SUPERVISION) \bowtie_{Employee=Number} (s_{Salary£40}(EMPLOYEES)))))$
(3.5)

The aftereffect of this articulation on the database in Figure 44 is appeared in Figure 49.

Number	Name
301	Steve Smith
375	Mary Smith

Equivalence of Mathematical Articulations

Relational algebra, in the same way as other formal languages, permits the detailing of articulations identical among themselves, which creates the same result. For instance, the accompanying comparability is legitimate where x, y, and z are real numbers:

$x \times (y + z) \,^{o}\, x \times y + x \times z$

For each value substituted for the three factors, the two articulations give a similar outcome. In relational algebra, we can give a comparative definition. A first thought of identicalness alludes to the database outline:

- $E1 \equiv_R E2$ if $E1(r) = E2(r)$, for each case of r in R.

Supreme identicalness is a more grounded property and is characterized as it takes after:

- $E1 \equiv E2$ if $E1 \equiv_R E2$, for each outline R.

The refinement between the two cases is because of the way that the properties of the operands are not indicated in the articulations (especially in the normal join operations). A case of supreme comparability is the following:

$p_{AB}(s_{A>0}(R) \,^{o}\, s_{A>0}(p_{AB}(R))$

while the accompanying proportionality

$$p_{AB}(R_1) \bowtie p_{AB}(R_2) \;_{R}p_{ABc}(R_1 \bowtie R_2)$$

holds just if in the mapping R the crossing point between the arrangements of traits of R_1 and R_2 is equivalent to A. Truth be told, if there were likewise different traits, the join would work just on A in the main articulation and on A and such other characteristics in the second, with various outcomes in general. The comparability of articulations in algebra is especially critical in inquiry advancement, which we talk about in Chapter 4. Indeed, SQL queries are converted into relational algebra, and the cost is assessed, cost being characterized as far as the extent of the middle of the road and last outcome.

At the point when there are diverse proportionate articulations, the one with the least cost is chosen. In this unique situation, comparability changes are utilized as well as operations that substitute one articulation for another comparable one. In specific, we are keen on those changes that can diminish the measure of the middle of the road relations or to set up an articulation for the use of one of the above changes. Let us delineate a first arrangement of changes.

- Atomization of determination, a conjunctive choice can be substituted by a course of nuclear determinations:

$$\sigma_{F_1 \wedge F_2}(E) \equiv \sigma_{F_1}(\sigma F_2(E))$$

where E is any articulation. This change permits the utilization of consequent changes that work on determinations with nuclear conditions.

- Falling projections, a projection can be changed into a course of projections that "dispose of" properties in different stages:

$$\pi_X(E) \equiv \pi_X(\pi_{XY}(E))$$

on the off chance that E is characterized on an arrangement of properties that contain Y (and X). This too is a preparatory change that will be trailed by others.

- Suspicion of the choice as for the join (frequently depicted as 'driving determinations down'):

$$s_F(E_1 \bowtie E_2) \,^{o}\, E_1 \bowtie s_F(E_2)$$

in the event that the condition F alludes just to qualities in the sub-articulation E_2.

- Expectation of the projection as for the join ('pushing projections down'). Let E_1 and E_2 be characterized on X_1 and X_2 individually, on the off chance that $Y_2 \subset X_2$ and $Y_2 \supseteq X_1 \cap X_2$ (so the properties in $X_2 - Y_2$ are not included in the join), at that point the following holds:

$$\pi_{X_1Y_2}(E_1 \bowtie E_2) \circ E_1 \bowtie \pi_{Y_2}(E_2)$$

By joining this rule with that of falling projections, we can get the accompanying comparability for theta-joins:

$$p_Y(E_1 \bowtie_F E_2)\ p_Y(p_{Y1}(E_1) \bowtie_F p_{Y2}(E_2)$$

where X_1 and X_2 speak to the traits of E_1 and E_2 separately, and J_1 and J_2 the particular subsets engaged with the join condition F, and at long last:

$$Y_1 = (X_1 \cap Y) \cup J_1$$
$$Y_2 = (X_2 \cap Y) \cup J_2$$

On the premise of the equivalences above, we can wipe out from each connection every one of the traits that do not show up in the last outcome and are not associated with the join.

- Blend of a determination and a Cartesian item to create a theta-join:

$$s_F(E_1 \bowtie E_2) \circ E_1 \bowtie_F (E_2)$$

Give us a chance to take a look at an illustration that illuminates the utilization of preparatory changes and the vital run of expectation of determinations. Let us assume that we wish to discover by alluding to the database in Figure 44 the enlistment quantities of the managers of the representatives more youthful than 30. A first articulation for this could be the particular of the Cartesian result of the two relations (which have no characteristics in like manner) trailed by a choice and at that point a projection:

$$p_{Head}((s_{Number=Employee\ Age<30}(\text{ EMPLOYEES} \bowtie \text{SUPERVISION}))$$

By methods for the past guidelines, we can essentially enhance the nature of this articulation, which is low for sure: it initially figures an expansive Cartesian item, in spite of the fact that the last outcome contains just a couple of tuples. Utilizing Rule 1, we separate the choice:

$$p_{Head}((s_{Number=Employee}(s_{Age<30}(\text{ EMPLOYEES} \bowtie \text{SUPERVISION}))$$

Furthermore, we would then be able to consolidate the main determination with the Cartesian item, frame an equi-join (Rule 5), and expect the second determination as for the join (Rule 3), getting:

$p_{Head}(s_{Age<30}($ EMPLOYEES) $\bowtie_{Number=Employee}$ SUPERVISION)

At last, we can take out from the main contention of the join, the pointless traits, utilizing Rule 4:

$p_{Head}(p_{Number}(s_{Age<30}($ EMPLOYEES) $\bowtie_{Number=Employee}$ SUPERVISION)

Some different changes can be valuable, especially different types of suspicion of choices and projections.

- Dispersion of the choice regarding the union:

$$\sigma_F(E_1 \cup E_2) \equiv \sigma_F(E_1) \cup \sigma_F(E_2)$$

- Dispersion of the choice regarding the distinction:

$$\sigma_F(E_1 \cup E_2) \equiv \sigma_F(E_1) - \sigma_F(E_2)$$

- Dispersion of the projection regarding the union:

$$\pi_X(E_1 \cup E_2) \equiv \pi_X(E_1) \cup \pi_X(E_2)$$

It is significant that projection is not distributive regarding distinction, as we can confirm by applying the articulations:

$$\pi_A(R_1 - R_2) \ \ and \ \ \pi_A(R_1) - \pi_A(R_2)$$

to two relations on A-B that contain tuples level with on A and diverse on B. Other fascinating changes are those in light of correspondence between set operations and complex determinations:

- $\sigma_{F_1 \vee F_2}(R) \equiv \sigma_{F_1}(R) \cup \sigma_{F_2}(R)$
- $\sigma_{F_1 \wedge F_2}(R) \equiv \sigma_{F_1}(R) \cap \sigma_{F_2}(R) \equiv \sigma_{F_1}(R) \bowtie \sigma_{F_2}(R)$
- $\sigma_{F_1 \wedge \neg(F_2)}(R) \equiv \sigma_{F_1}(R) - \sigma_{F_2}(R)$

At that point, there is the commutative and acquainted property of all the parallel operations barring contrast and the distributive property of this join with regard to the union:

E$\bowtie$ (E_1 Ç E_2) ° (E$\bowtie$ E_1) Ç (E$\bowtie$ E_2)

At last, we ought to know that the presence of intermediate outcomes (relations with the zero tuples) makes it conceivable to rearrange articulations normally. Note that a join (or additionally a Cartesian item) in which one of the operations is the void connection, creates an unfilled outcome.

Algebra with Null Values

In the above areas, we have constantly underestimated that the mathematical articulations were being connected to relations containing no invalid esteems-values. In Section 2.1.5, we talked about the significance of invalid esteems-values in real applications. We should in any event touch upon the effect that they have on the languages talked about in this section. The problem is managed further and explained in Chapter 4, with regards to the SQL language.

Let us look at the connection in Figure 50 and the accompanying choice:

EMPLOYEES

Number	Name	Age	Salary
101	Mary Smith	34	40
103	Mary Bianchi	23	35
104	Luigi Neri	38	61
105	Nico Bini	44	38
210	Marco Celii	49	60
231	Siro Bisi	50	60
252	Nico Bini	44	70
301	Steve Smith	34	70
375	Mary Smith	50	65

SUPERVISION

Head	Employee
210	101
210	103
210	104
231	105
301	210
301	231
375	252

Figure 50 – Null Values Relation.

(Source: Paredaens, J. D. The Structure of the Relational Database)

Presently, the primary tuple of the connection must add to the outcome and the second should not, rather what would we be able to say in regards to the third? Naturally, the age value is an invalid of obscure type or sort, in that the value exists for every individual, and the invalid implies that we disregard it. As for these queries, rather than the customary two-valued logic (in which equations are either true or false) a three-valued logic can be utilized. In this logic, a recipe can be valid or false or can accept a third, new truth value that we call obscure and speak to by the symbol U. A nuclear condition accepts this value when no less than one of the terms of the correlation accepts the invalid value. In this manner, the primary tuple absolutely has a place with the outcome (true), the second positively does not have a place (false) and the third maybe has a place and maybe does not (obscure). The determination creates therefore the tuples for which the equation is valid. The semantic premise of the three connectives is the possibility that the obscure value is somewhere close to true and false.

not	
F	T
U	U
T	F

and	T	U	F
T	T	U	F
U	U	U	F
F	F	F	F

or	T	U	F
T	T	T	T
U	T	U	U
F	T	U	F

We should bring up that the three-valued logic for logarithmic administrators additionally shows some unsuitable properties. For instance, let us consider the algebraic articulation

$s_{Age>30}$(PEOPLE) Ç $s_{Age\ 30}$(PEOPLE)

Sensibly, this articulation should return accurately the PEOPLE connection, given that the age value is either higher than 30 (first sub-articulation) or is definitely not higher than 30 (second sub-articulation). Then again, if the two sub-expressions are assessed independently, the third tuple of the case (simply like some other tuple with an invalid an incentive for Age) has an obscure outcome for each sub-articulation and for the union. Just by methods for a worldwide assessment (unquestionably unreasonable on account of complex articulations) would we be able to land at the decision that such a tuple should positively show up in the outcome. The same goes for the articulation

$s_{Age>30ÚAge\ 30}$(PEOPLE)

in which the dis-junction is assessed by the three-valued logic. The best technique for defeating the troubles portrayed previously is to treat the invalid values from an absolutely syntactic perspective. This approach works similarly for both two-valued logic and three-valued logic. Two new types of nuclear states of determination are acquainted checking whether a value is indicated or invalid:

- A is invalid expecting the value valid on a tuple t if the estimation of t on A is invalid and false if it is not; and

- A is not invalid expecting the esteem valid on a tuple t if the estimation of t on A originates from the area of A and false if the esteem is invalid.

In this specific circumstance, the articulation

$s_{\text{Age>30}}$ (PEOPLE)

restores the persons whose age is known and more than 30, while to acquire those who are or could be more than 30 (which is, those whose age is actually known and more than 30 or, on the other hand not known), we can utilize the articulation:

$s_{\text{Age>30ÚAge IS NULL}}$ (PEOPLE)

$s_{\text{Age>30}}$(PEOPLE) Ç $s_{\text{Age 30}}$(PEOPLE)

$s_{\text{Age>30Ú Age 30}}$ (PEOPLE)

Additionally, the articulations trying not to restore a whole connection but just the tuples that have a value not invalid for Age. On the off chance that we need the whole connection as the outcome, we need to include an 'is invalid' condition:

$s_{\text{Age>30ÚAge 30£ÚAge IS NULL}}$ (PEOPLE)

This approach is utilized as a part of the present rendition of SQL, which underpins a three-valued logic and is usable in prior variants, which embraced a two-value logic.

Views

In Chapter 1, we perceived how it can be helpful to make diverse portrayals of similar information accessible to clients. In the relational model, this is accomplished by methods for determined relations, that is, relations whose substance is characterized in terms of the substance of different relations. In a relational database, there exists base relation in which the substance is independent and really put away in the database, and inferred relation in which the substance is gotten from the substance of different relations. It is conceivable that a determined connection is characterized in terms of other inferred relations, on condition that a request exists among the inferred relations, with the goal that every determined connection can be communicated in wording of base relations.

There are fundamentally two sorts of inferred relations.

- Appeared-materialized sees (views) that determine relations that are really put away in the database.
- Virtual relations (additionally called views) referred to as relations characterized by methods for capacities (articulations in the question-query language), not put away in the database but usable in the queries.

Emerged views have the up-side of being quickly accessible for queries. Often, it is a substantial undertaking to keep up their substance steady with those relations from which actually they are inferred, as any of the change to the real base-relations from which they actually depend must be spread to them. Then again, virtual relations must be recalculated for each query; however, this delivers no consistency issues. Generally, we can state that appeared views are advantageous when there are less updates than queries and the count of the view is complex. It is troublesome to give general methods for keeping up consistency between base-relations and materialized-views. Therefore, most business systems give instruments for sorting out just virtual relations, which from now on, with no danger of equivocalness, we call essential views (Ceri S. G., 1989).

Views are characterized in relational systems by methods of query language articulations. At that point, inquiries on views are settled by substituting the meaning of the view for the view itself, that is, by forming the first inquiry with the view query. For instance, considering a database on this relation:

$R_1(ABC), R_2(DEF), R_3(GH)$

with a view characterized utilizing a Cartesian item took after by a choice, on this pattern, the query

$R = s_{A>D}(R_1 \bowtie R_2)$

is executed by supplanting R with its definition

$s_{B=G}(R_1 \bowtie R_3)$

The utilization of perspectives can be helpful for an assortment of reasons.

- A client inspired by just a segment of the database can abstain from managing the superfluous parts. For instance, in a database with two relations on the patterns

EMPLOYEES (Employee, Department)

MANAGERS (Department, Supervisor)

a client who is intrigued just in the employees and their separate managers could discover his undertaking encouraged by a view characterized as taken after:

$p_{Employee,Supervisor}$(EMPLOYEES $\bowtie$ MANAGERS)

- Very mind boggling articulations can be characterized utilizing views, with specific focal points on account of rehashed sub-articulations.
- By methods to get approvals related with views, we can present instruments for the security of protection. For example, a client could be conceded limited access to the database through a particularly outlined view.
- In case of rebuilding a database, it can be helpful to characterize views comparing to relations that are never again display after the rebuilding. Along these lines, applications composed with reference to the prior rendition of the blueprint can be utilized on the better and brighter one without the requirement for changes. For instance, if a diagram like R(ABC) is supplanted by two diagrams R_1 (AB), R_2 (BC), we can characterize a view, $R = R_1 \bowtie R_2$ and leave in place the applications that allude to R. If B is actually a key for R_2, the nearness of the view is totally straightforward.

Given a refresh on a view, we might want to have precisely one arrangement of updates to the base relations, with the end goal that the view, if registered after these progressions to the base relations, shows up as though the given refresh had been performed on it. Shockingly, this is not for the most part conceivable. For instance, let us look once more at the view

$p_{Employee,Supervisor}$(EMPLOYEES $\bowtie$ MANAGERS)

Expecting that we need to embed a tuple into the view, we might want to have tuples to embed into the base relations that permit the era of the new tuple in the view. In any case, this is impractical, in light of the fact that the tuple in the view does not include the Department trait, thus we do not have an incentive for it, as required to set up the correspondence between the two relations.

All in all, the issue of refreshing perspectives is mind boggling. All systems have solid constraints in regards to the refreshing of perspectives.

RELATIONAL ALGEBRA

The term relational algebra alludes to a group of query language in view of the first place arrange predicate analytics. These are described by being explanatory, implying that the inquiry is indicated as far as the property of the outcome, as opposed to the method to be followed to acquire it. Relational algebra is actually known as a procedural type of language, since its looks specific (by methods for the individual uses of the administrators) and shows the development of the outcome which is well ordered.

There are numerous forms of relational algebra and it is impractical to introduce them all here. We initially outline the adaptation that is closest to predicate math, domain relational analytics that shows the fundamental attributes of these languages. We at that point talk about the confinements and changes that make it of viable intrigue and interest. We will subsequently present tuple analytics with run affirmations, which shapes the reason for a large number of the developments accessible for inquiries in SQL.

With regards to the subjects as of now examined concerning the relational display, we utilize non-positional documentation for relational algebra. This area (on math) and the accompanying one (on Datalog) can be discarded without disabling the comprehension of the other parts of the book. It is not important to be familiar with the first request predicate analytics in request to peruse this area. We give now a few remarks that empower anybody with earlier learning to get a handle on the association with the first request predicate math. These remarks might be precluded without trading off the comprehension of ensuing ideas. There are a few rearrangements and adjustments in relational algebra regarding the first request predicate analytics. In the first place, in predicate analytics, we have predicate images-symbols (deciphered similarly as relations) and work-function images (deciphered as capacities). In relational algebra, the predicate images are compared to relations in the database (aside from other standard predicates, for example, fairness and disparity) and there are no work-function images. (They are not required given the level structure of the relations.)

At that point, in predicate math, both open equations (those with free factors) and shut equations (those whose factors are altogether bounded and none of them are free) are of interest. The second sort have a reality value that, as for an elucidation, is settled while the first have a value that relies upon the values substituted for the free factors. In relational analytics, just the open equations are of actual interest. A query is characterized by methods for an open analytics equation and the outcome comprises of tuples of qualities that fulfill the recipe at the point when substituted with the expectation of complimentary factors (Abiteboul, 1995).

Domain-Area Relational Analytics

Relational analytics articulations have this frame:

$\{A_1{:}x_1,\ldots, A_k{:}x_k | f\}$

where:

- $A_1, \ldots, A_k$ are unmistakable characteristics (which do not really need to show up in the blueprint of the database on which the question is figured);
- $x_1, \ldots, x_k$ are factors (which we will take to be particular for accommodation, regardless of the possibility that this is not entirely essential); and
- f is a recipe, as indicated by the accompanying principles.
- There are two sorts of nuclear recipe:

* $R(A_1{:}\, x_1, \ldots, A_p{:}x_p)$, where $R(A_1, \ldots, A_p)$ is a social blueprint and also, $x_1, \ldots, x_p$ are factors; and

* $x\vartheta y$ or even $x\vartheta c$, with x and also y factors, c consistent and ϑ examination administrator ($=, \neq, \leq, \geq, >, <$).

- If f_1 and f_2 are recipes, at that point $f_1 \vee f_2$, $f_1 \wedge f_2$, and $\neg f_1$ are equations ($\vee$, $\wedge$, $\neg$ are the coherent connectives); where important, with a specific end goal to guarantee that the priorities are unambiguous, sections can be utilized; and

- if f is a recipe and x a variable (which generally shows up in f, regardless of the possibility that not entirely essential), ∃x(f) and ∀x(f) are equations (∃ and ∀ are the existential quantifier and all-inclusive quantifier, separately).

The rundown of sets $A_1 : x_1, \ldots , A_k : x_k$ is known as the objective rundown since it characterizes the structure of the outcome, which is comprised of the connection on $A_1, \ldots , A_k$ that contains the tuples whose esteems (values) when substituted for $x_1, \ldots , x_k$ render the recipe genuine or differently stated as true. The formal meaning of reality estimation of an equation goes past the extent of this book and in the meantime its significance can be clarified. Let us quickly revise the syntactic structure of equations (the expression "esteem" "value" here signifies 'a component of the area', where we expect, for effortlessness, that all properties have a similar space):

- a nuclear equation $R(A_1{:}x_1, \ldots , A_p{:}x_p)$ is valid for estimations of $x_1, \ldots , x_p$ that shape a tuple of R;
- a nuclear equation $x\vartheta y$ is valid for estimations of x and y with the end goal that the estimation of x remains in connection ϑ with the estimation of y; also for $x\vartheta c$;
- the importance of connectives is the typical one; and
- for the equations worked with quantifiers:
    - ∃x(f) is valid if there exists no less than one incentive for x that makes f genuine;
    - ∀x(f) is valid if f is valid for every single conceivable incentive for x.

Give us now a chance to delineate relational algebra by demonstrating how it can be utilized to express the queries over the composition:

$$Employees\left(\underline{Number},\ Name,\ Age,\ Salary\right)$$

$$Supervision\left(Head,\ \underline{Employee}\right)$$

Let us start with an exceptionally basic inquiry: discover the enrolment numbers, names, ages, and pay rates of the workers acquiring > than 40,000, which we can figure in relational algebra with a determination:

$$s_{Salary>40}(\text{EMPLOYEES}) \qquad (3.6)$$

There is a similarly basic detailing in relational algebra, with the articulation:

{ Number: m, Name: n, Age:a, Salary: s |

EMPLOYEES(Number:m, Name: n, Age:a, Salary : s) • s > 40}
(3.7)

Note the nearness of two conditions in the equation (associated by the legitimate administrator and):

- the first, EMPLOYEES(Number:m,Name:n,Age:a,Salary:s), requires that the values substituted separately for the factors m, n, a, s constitutes with a tuple of the connection EMPLOYEES; and
- the second requires that the estimation of the variable s is more prominent than 40.

The outcome is comprised of the estimations of the four factors that begin from the tuples of the area EMPLOYEES for which is actually the estimation of the compensation is more noteworthy than 40 thousand.

A marginally more unpredictable inquiry is: discover the enlistment numbers, names, and periods of the workers who gain > than 40 thousand. This inquiry requires a subset of the traits of EMPLOYEES and along these lines in polynomial math can be figured with a projection (Formula 3.1):

$p_{Number,Name,Age}(s_{Salary>40}(EMPLOYEES))$

This inquiry in analytics can be defined in different ways. The most immediate, if not the least complex, depends on the perception that what intrigues us are the estimations of Number, Name, and Age, which shape some portion of the tuples for which compensation is more than 40. That actually is for those whose estimation-value of Salary is greater than 40, which permits the culmination of a tuple of the connection Representatives-EMPLOYEES. We would thus be able to utilize an existential quantifier:

{Number: m, Name : n, Age:a

$s(EMPLOYEES(Number: m, Name : n, Age: a, Salary : s) • s > 40)
(3.8)

The utilization of the quantifier is not really vital, since by basically composing

{Number: m, Name : n, Age:a |

EMPLOYEES(Number: m, Name : n, Age: a, Salary : s) • s > 40) (3.9)

we can get a similar outcome.

A similar structure can be stretched out to more mind boggling inquiries, which in relational algebra we planned utilizing the join operation. We will require more nuclear conditions, one for every connection included, and then we can utilize rehashed factors to demonstrate the join conditions. For instance, the question that demands the enlistment quantities of the supervisors-administrators of the employees who earns more than 40 thousand, detailed in variable based math by Formula 3.2:

$p_{Head}(SUPERVISION \bowtie_{Employee=Number}(s_{Salary>40}(Employees)))$

can be detailed in analytics by:

{Head:h| EMPLOYEES(Number: m, Name : n, Age: a, Salary : s) •
SUPERVISION (Employee:m, Head:h) • s > 40} (3.10)

where the variable m, basic to both nuclear conditions, assembles the same correspondence between tuples indicated in the join. Here, likewise, we can utilize existential quantifiers for every one of the factors that do not show up in the objective rundown. As for the situation over, this is redundant and would muddle the plan. In the event that the inclusion of various tuples of a similar connection is required in an articulation, it is adequate to incorporate more limitations on the same predicate in the recipe with various factors. Consider the inquiry: discover the names and pay rates of the heads of the employees earning more than 40 thousand, communicated in algebra by Formula 3.3, which has a join of the connection with itself:

$p_{NameH,SalaryH} (Q_{NumberH,NameH,SalaryH,AgeH\ Number,Name,Salary,Age} (EMPLOYEES)$
$\bowtie_{NumberH=Head}$
$(SUPERVISION \bowtie_{Employee=Number} (s_{Salary>40}(EMPLOYEES))))$

This question is detailed in analytics by requiring, for each one of the tuples of the result, the presence of three tuples: first, identify the employees earning more than 40 thousand; second, determine their corresponding heads in the SUPERVISION connection; and third, retrieve the details of those heads in the EMPLOYEES connection.

{NameH : nh, SalaryH:sh |

EMPLOYEES (Number: m, Name: n, Age: a, Salary: s) ^ s > 40 ^

SUPERVISION(Employee:m, Head : h) ^

EMPLOYEES(Number h, Name: nh, Age : ah, Salary: sh) } (3.11)

Consider next the query: discover the employees earning more than their respective heads, indicating enlistment number, name, and pay of the relevant employees and heads (Formula 3.4). This varies from the previous one, in the need of contrasting estimations-values of the same quality beginning from various tuples, which causes no specific issues:

{Number : m, Name : n, Salary : s, NumberH : h, NameH : m, Salary : sh |

EMPLOYEES (Number: m, Name: n, Age: a, Salary: s) ^

SUPERVISION(Employee:m, Head : h) ^

EMPLOYEES(Number h, Name: nh, Age : ah, Salary: sh) ^ s > sh}
(3.12)

The last case requires a more intricate arrangement. We should discover the enlistment numbers and names of the heads whose employees earn more than 40 thousand. In variable based math we utilized a distinction (Formula 3.5) that creates the required set by considering every one of the directors aside from the individuals who have no less than one employee earning not exactly 40 thousand:

$p_{\text{Number, Name}}$(EMPLOYEES $\bowtie_{\text{Number=Head}}$
(p_{Head}(SUPERVISION) —
p_{Head}(SUPERVISION $\bowtie_{\text{Employee=Number}}$ ($s_{\text{Salary£40}}$(EMPLOYEES)))))

In analytics, we should utilize a quantifier. By making an indistinguishable stride in algebra, we can utilize a nullified existential quantifier. We utilize a significant number of these, one for every variable included.

{Number: h, Name:n | EMPLOYEES(Number:h, Name: n, Age: a, Salary: s) ^

SUPERVISION(Employee:m, Head:h) ^

—$m'($ n'($s'EMPLOYEES(Number: m', Name: n', Age: a', Salary: s') ^

SUPERVISION(Employee:m', Head:h) ^ s' £ 40))))} (3.13)

As an option, we can utilize all-inclusive quantifiers:

{ Number : h, Name : n | EMPLOYEES(Number : h, Name : n, Age :a, Salary : s) ∧

SUPERVISION(Employee:m, Head : h) ^

" m'(" n'(" 'a(Ø(Employees (Number: m', Name : n', Age : a', Salary: s') ^

SUPERVISION(Employee : m', Head : h)) Ú s' > 40))))}
(3.14)

This articulation chooses an administrator h if for each fourfold of qualities m′, n′, a′, s′ in respect to the representatives of h, s′ is more than 40. The structure ¬fVg compares to the condition 'If f then g' (for our situation, if m′ is an worker having h as an administrator, at that point the compensation of m′ is more prominent than 40), given that it is valid in all cases separated from the one in which the f is valid and g is false. It is important that varieties of de Morgan laws legitimate for boolean variable based math administrators, with the end goal that:

Ø(f ^ g) ° Ø(f) Ú (g)

Ø(f Ú g) = Ø(f) ^ Ø (g)

$x(f) = Ø("xØ(f)))

("x(f) = Ø$x(Ø(f)))

Pros and Cons of Domain Analytics

As we have illustrated, relational algebra has some few intriguing viewpoints, especially its decisive nature. There are, notwithstanding, a few deformities and constraints that are critical from the viable point of view.

To begin with, take note that analytics permits articulations that do not have a real reason. For a case, the articulation:

$$\{A_1 : x_1, A_2 : x_2 \mid R(A_1 : x_1) \wedge x_2 = x_2\}$$

creates subsequently a connection on A_1 and A_2 made up of tuples whose values in A_1 show up in the connection R, and the incentive on A2 is any estimation of the area (since the condition $x_2 = x_2$ is constantly valid). Specifically, if the space changes, for instance, from the whole numbers within 0 to 99 to the whole numbers amongst 0 and 999, the response to the inquiry additionally changes. On the off chance that the area is limitless, the appropriate response is additionally vast, which is undesirable. A comparative perception can be made for the articulation

$\{A_1 : x_1 \mid R(A_1 : x_1))\}$

the aftereffect of which contains the estimations-values of the area not showing up in R.

It is helpful to present the accompanying idea here. A statement of a query language is area-domain autonomous if its outcome, on each example of the database, does not fluctuate on the off chance that we change the space on the premise of which the articulation is assessed. A language is area-domain autonomous if all its articulations are area autonomous. The necessity of space freedom is obviously major for genuine language, since space subordinate articulations have no down to earth-practical utilization.

In light of the articulations seen above, we can state that relational algebra is not space-domain autonomous. In the meantime, it is difficult to see that relational algebra is actually domain-area free since it develops the outcomes from the relations in the database, while never alluding to the areas-domains of the characteristics. So the estimations-values of the outcomes all originate from the example to which the articulation is connected. In the event that two inquiry languages are proportional when for every articulation in one of the languages there exists a proportionate articulation in the other language, and the other way around, we can express that polynomial math and analytics are not equal. This is on the grounds that calculus, not at all like algebra, permits articulations that are domain-area subordinate.

In the event that we confine our thoughtfulness regarding the subset of algebra made up exclusively of articulations that are space-domain autonomous, we get a language that is without a doubt identical to relational algebra. Actually,

- for each declaration of relational analytics that is area autonomous there exists a declaration of relational algebra identical to it; and
- for each declaration of relational algebra there is a statement of analytics identical to it (and consequently space autonomous).

The verification of identicalness is beyond the content of this book. However, we can say its essential standards. There is a correspondence between choices and basic conditions, among projection and existential evaluation, among join and conjunction, among union and disjunction, and between distinction and conjunction related with nullification. The general quantifiers can be overlooked in that they could be changed to existential-quantifiers utilizing de Morgan's laws (Paolo, Stefano, Stefano, & Riccardo, 1999).

Relational calculus has another weakness, that of requiring various factors, frequently one for each quality of every connection included. At that point, when evaluations are essential the quantifiers are likewise duplicated. The main practical language construct at any rate to a limited extent with respect to space analytics, known as Query-by-Example (QBE), utilizes a realistic interface that liberates the client from the need to determine dreary subtle elements.

Keeping in mind the end goal to beat the confinements of area-domain analytics, a variation of relational analytics has been proposed, in which the factors mean tuples rather than single esteems-values. Along these lines, the quantity of factors is regularly diminished, in that there is just a variable for every connection included. This tuple would however be proportional to domain analytics, and in this manner have the restriction of space reliance.

Subsequently, we like to overlook the introduction of this language. Rather we will move straightforwardly to a language that has the qualities of tuple math, furthermore, in the meantime beats the deformity of space-domain reliance, by utilizing the immediate relationship of factors with relations of the database. The following segment manages this language.

Tuple Math with Range Revelations

The statements of tuple analytics with go affirmations have the shape

$\{T|L|f\}$

where:

L- is the range list, counting the free factors of the equation f, with the particular scopes of changeability. Actually, it is a rundown of components of sort x(R), with x variable and R connection name; if x(R) is in the range list, at the point when the articulation is assessed, the conceivable esteems-values for x are quite recently the tuples in the connection R;

T - is the objective rundown, made out of components of sort Y : x.Z (or just x.Z, contraction for Z : x.Z), with x-variable and Y plus Z arrangements of characteristics (of equivalent length); the properties in Z must show up in the mapping of the connection that makes up the scope of x. We can likewise compose x. *, as shortened form for X : x.X, where the scope of the variable x is a connection on properties X;

f - is an equation with

- particles of sort $x.A\vartheta c$ or $x_1.A_1\ \vartheta\ x_2.A_2$, which think about, individually, the estimation of x on the characteristic A with the consistent c and the estimation of x_1 on the A_1 with that from x_2 on the A_2;
- connectives with respect to space analytics;
- quantifiers, which additionally relate reaches to the particular factors:

$x(R)(f) "x(R)(f)

where, $\exists x(R)(f)$ signifies 'there is a tuple x in the connection R that fulfills the equation f' and $\forall x (R)(f)$ signifies 'each tuple x in R fulfills f'.

Range affirmations in the range list and in the measurements have an imperative part. While presenting a variable x, a range assertion R(x) determines that x can accept as qualities just the tuples of the connection R with which it is related. Subsequently, this language has no need of nuclear conditions, for example, those found in area-domain analytics, which indicate that a tuple has a place with a connection.

Next we show how the different inquiries that we have effectively communicated in algebra and area-domain analytics can be detailed in this language. The first query, which demands enrolment number, name, age, and pay rate of the employees earning more than 40 thousand, turns out to be extremely compact and clear (contrast and Formuna 3.7):

{e.* |e(EMPLOYEES) |e.Salary>40} (3.15)

Keeping in mind the end goal to deliver just a portion of the properties, enlistment number, name, and age of the employees earning more than 40 thousand (Articulation 3.1 in variable based math and Formula 3.9 in domain analytics), it is adequate to alter the objective rundown:

{e.(Number, Name, Age) |e(Employees)} | e.Salary>40} (3.16)

For inquiries including more than one connection, more factors are essential, determining the states of relationship on the qualities. The question that demands discover the enlistment quantities of the heads of the employees gaining more than 40 thousand (Formula 3.1 in domain analytics) can be figured with:

{s.Head}|e(Employees), s(SUPERVISION) |

e.Number = s.Employee ^ e.Salary>40} (3.17)

Note how the recipe takes into account the conjunction of two nuclear conditions: one that compares to the join condition (e.Number=s. Employee) and the other join to the typical determination condition (e.Salary >40). On account of articulations that compare to the join of a connection with itself, there will be more factors with a similar range. The query: find names and compensations of the heads of the employees earning more than 40 thousand (Formula 3.3 and Formula 3.11) can be defined utilizing the accompanying articulation:

{NameH, SalaryH : e›.(Name, Salary) |

e'(EMPLOYEES), s(SUPERVISION), e(EMPLOYEES) |

e'.Number = s.Head ^ s.Employee = e.Number ^

e.Salary > 40} (3.18)

Also, we can discover the employees who earn more than their corresponding heads, indicating enlistment number, name, and salary of the employees and heads (Formula 3.4 in relational algebra and Formula 3.12 in domain analytics):

{e.(Name, Number, Salary), NameH, NumberH, SalaryH : e'.(Name, Number, Salary) |

e(EMPLOYEES), s(SUPERVISION), e' (EMPLOYEES) |

e. Number = s.Employee ^ s.Head = e› Number ^

e. Salary > e›. Salary (3.19)

Queries with quantifiers are considerably more brief and reasonable here than in area math. The inquiry that retrieves the enrolment number and name of the heads whose employees whose salary is greater than 40 thousand (Formula 3.5 in algebra and Formula 3.13 or Formula 3.14 in domain analytics) can be communicated with far less quantifiers and factors. Once more, there are different choices, in light of the utilization of the two quantifiers and of refutation. With all-inclusive quantifiers:

{e.(Number, Name) | e(EMPLOYEES), s(SUPERVISION) |

e.Number = s.Head ^ "e'(EMPLOYEES)("s'SUPERVISION)

(Ø (s.Head = s'.Head ^ s›.Employee = e›.Number) Ú

e›.Salary > 40))} (3.20)

With refuted existential quantifiers:

{e.(Number, Name) | e(EMPLOYEES), s(SUPERVISION) |

e.Number = s.Head ^ Ø($e'(EMPLOYEES)($s' (SUPERVISION)

(s.Head = s''.Head ^ s'.Employee = e'.Number ^

e'.Salary £ 40)))} (3.21)

For unknown reasons, it is impractical in tuple math with extend statements to express every one of the queries that could be figured in relational algebra (or in domain-area analytics). Specifically, the inquiries in relational-algebra, which require the union operation, cannot be communicated in this rendition of math. Take for instance, the basic union of two relations on the same traits: given R_1 (AB) and R_2 (AB), we wish to detail the inquiry that would express in variable based math by the union of the R_1 and the R_2. On the off chance that the articulation had two free factors, each tuple of the outcome would need to compare to the tuple of each one of the relations. This is a bit much, on the grounds that the union requires the tuples of the outcome to show up in no less than one of the operands, not fundamentally in both. On the off chance that the articulation had a solitary free variable, this would need to allude to a solitary connection without gaining tuples from the other for the outcome. In this manner, the union cannot be communicated.

Thus, SQL takes into account an express union develop to express queries that would somehow demonstrate unthinkable. This is on the account of the explanatory parts of SQL depend on tuple analytics with extend affirmations.

Note that on the off chance that we permitted the meaning of reaches made up of at least two relations, we would resolve the issue of straightforward unions. We proved unable nonetheless, define complex unions whose operands are sub-articulations not specifically comparing to connection blueprints. For instance, given two relations R_1 (ABC) and R_2 (BCD), the union of the projections on BC

$p_{BC}(R_1)$ Ç $p_{BC}(R_2)$

could not be communicated even with this expansion, on the grounds that the two relations have distinctive plans, and in this way a solitary variable cannot be related with both.

We should notice that, while the union operation cannot be communicated in this adaptation of analytics, the crossing point and contrast operations are expressible.

- Intersection requires that the tuples of the actual outcome have a place with both the operands and consequently the outcome can be built with reference to simply one connection, with the extra condition that requires the presence of an equivalent tuple in the other connection. For instance, the crossing point:

$p_{BC}(R_1)$ $p_{BC}(R_2)$

can be communicated by:

$\{x_1.BC|x_1(R_1)$ |$\$x_2(R_2)(x_1.B=x_2B\text{^}x_1.C=x_2.C\}$

- Similarly, the distinction that creates the tuples of an operand that are not contained in the other, can actually be determined by asking correctly those tuples of the first contention that do not show up in the second. For illustration,

$p_{BC}(R_1)$ - $p_{BC}(R_2)$

$\{x_1.BC|x_1(R_1)$ |Ø$\$x_2(R_2)(x_1.B=x_2B\text{^}x_1.C=x_2.C\}$

DATALOG

We finish up this section with a short summary of another database query language that has produced impressive interest for mainstream researchers since the mid-eighties. The fundamental idea on which the actual Datalog language is based, is that of adjusting the rationale programming language 'Prolog' for use with the databases. We cannot show either Datalog or Prolog in detail here. However, we can say the most intriguing angles especially from the perspective of a correlation with alternate language found in this section.

In its essential form, Datalog is a disentangled rendition of Prolog, a language in light of first request predicate calculus, but with an alternate approach from the relational analytics previously mentioned. There are two sorts of predicate in Datalog:

- extensional predicates that compare to the relations in database; and
- intensional predicates that relate basically to views (virtual relations) indicated by methods for sensible standards.

Datalog rules have the frame:

head ˜ body

where

- the head is a nuclear recipe of the frame $R(A_1 : a_1, \ldots, A_p : a_p)$, like those utilized as a part of space social calculus, where every a_i, can be a consistent/variable;
- the body is a rundown of nuclear equations, of the two structures permitted in space analytics, that is, the frame R(…) and the correlation between factors or between a variable and a steady.

Guidelines characterize the "substance" of intentional predicates, as the tuples which values fulfill the body. The accompanying conditions are forced:

- extensional predicates can seem just in the collection of principles;
- if a variable shows up in the leader of an rule, it should likewise show up in the body of a similar rule; and
- if a variable shows up in an examination particle, it should likewise show up in an iota of the shape R(…) in a similar body.

The main condition guarantees that there will be no endeavour to rethink that the relations be put away in the database. The other two guarantee a property comparable (in this specific situation) to area autonomy as talked about with respect to relational algebra.

An essential normal for Datalog, which recognizes it from the other languages we have until now, is its utilization of recursion. It is feasible for an intentional predicate to be characterized as far as itself (straightforwardly or by implication).

We will come back to this perspective right away. Datalog queries are indicated basically by methods for iotas $R(A_1 : a_1, \dots , A_p : a_p)$. Ordinarily went before by a question mark '?' to underline definitely the way that they are inquired, but other syntactic traditions might be utilized. Queries create as results tuples of the connection R, which can be acquired by reasonable substitutions for the factors. For instance, the query:

?EMPLOYEES (Number : m, Name : n, Age : 30, Salary : s)

Presents the employees who are 30 years of age. To detail more mind boggling queries, we should utilize rules. For instance, with a goal to discover the enlistment quantities of the heads of the employees who earn more than 40 thousand, figured in relational by Expression 3.2 and in area analytics by Articulation 3.10, we characterize an intentional predicate SUPEROFRICH, running the same rule:

SUPEROFRICH(Head:h) ˜

EMPLOYEES(Number : m, Name: n, Age: a, Salary: s),

SUPERVISION(Employee:m, Head: h), s> 40 (3.22)

With a specific end goal to assess an inquiry of this nature, we should characterize the semantics of the guidelines. The essential idea is that the body of a run is considered as the conjunction of the iotas that show up in it, and in this way the run can be assessed similarly as a declaration of area math. The body of the articulation, substituting the commas and turns into the equation, and the leader of the articulation, aside from the name of the intentional predicate, turns into the objective rundown. Articulation 3.22 characterizes the intentional connection SUPEROFRICH as made up of the same tuples that show up in the consequence of Expression 3.10 of analytics, which has the structure portrayed above:

{Head: h | EMPLOYEES(Number:m, Name:n, Age: a, Salary: s) ^ SUPERVISION(Employee:m, Head: h) ^ s > 40}

Correspondingly, we can compose rules for a large number of the queries discussed in the previous segments. In the nonappearance of recursive definitions, the actual semantics of the Datalog is in this way exceptionally straightforward, in the feeling that the different intentional predicates can be figured by implies of articulations like calculus. Be that as it may, utilizing the definition given until this point for Datalog, it is impractical to figure every one of the inquiries that could be communicated in calculus (and in relational algebra). This is on the grounds that there is no accessible comparing to the all-inclusive quantifier (or to invalidation in the full feeling of the term). It can be demonstrated that non-recursive Datalog is identical to the space autonomous subset of math (calculus) without refutations or widespread quantifiers.

To outfit Datalog with an indistinguishable expressive power from calculus, we should add to the essential structure the likelihood of incorporating into the body, not only nuclear conditions but also refutations of nuclear conditions (which we show by the image-symbol not).

Just along these lines, would we be able to figure the query to determine the number and name of the heads whose employees earn more than 40 thousand (Expression 3.13):

{ Number: h, Name:n | EMPLOYEES(Number:h, Name:n, Age: a, Salary : s)

SUPERVISION(Employee : m, Head : h) ^

Ø$m'($n'($a'($s'(EMPLOYEES(Number: m', Name : n', Age: a', Salary: .s') ^

SUPERVISION(Employee:m›, Head : h) ^ s› £ 40))))}

Give us a chance to continue by characterizing a predicate for the heads who do not fulfil the condition:

SUPEROFSOMENOTRICH(Head: h) ˜

SUPERVISION(Employee:m, Head : h),

EMPLOYEES(Number : m, Name: n, Age: a, Salary:s), s’ £ 40

We can utilize this predicate in the discredited shape:

SUPEROFALLRICH(Number : h, Name : n) ˜

EMPLOYEES(Number : h, Name:n, Age: a, Salary: s)

SUPERVISION(Employee : m, Head : h),

NOT SUPEROFSOMENOTRICH(Head : h)

We could demonstrate that non-recursive Datalog with nullification is proportional to the space-domain free subset of calculus.

More prominent expressive power is acquired by utilizing recursive principles. For illustration, alluding again to the actual database with the relations EMPLOYEES and SUPERVISION, we can characterize that the intentional predicate the SUPERIORS, which gives, for every employee, the head, the super head, etc. without any points of confinement. For this we require two principles:

SUPERIORS(Employee : e, SuperHead : h) ˜

SUPERVISION(Employee : e, Head:h)

SUPERIORS(Employee:e, SuperHead : h) ˜

SUPERVISION(Employee:e, Head:h')

SUPERIORS(Employee : h', SuperHead : h)

The second manage is recursive, in that it characterizes the SUPERIORS connection in terms of itself. To assess this manage, we cannot continue as we have done up to now, in light of the fact that a solitary assessment of the body would not be adequate to compute the recursive predicate. There are different procedures for formally characterizing the semantics for this situation, yet they are beyond the scope of this material. We will touch upon the easiest strategy, in view of the procedure known as fix-point. The principles that characterize the intentional recursive predicate are assessed ordinarily until the point that a cycle does not produce new outcomes. In our case, the main emphasis would create a connection SUPERIORS equivalent to the extensional connection SUPERVISION that contains the heads of the employees. The second step would include the head (SuperHead) of the head. Clearly, questions of this nature cannot be planned in relational algebra (or in analytics) on the grounds that we would have no chance to get of knowing how often the join of the connection SUPERVISION with itself must be determined.

As a last issue before finishing up, we basically express the way that certain recursive principles with nullification are hard to assess, on the grounds that the fix-point cannot be attained. The user is ought to know that it is conceivable to distinguish a superbly workable subset of recursive Datalog with invalidation that is substantially more than the relational calculus and the relational algebra:

- for each statement of variable based math there is a proportionate articulation of Datalog with nullification; and
- there are recursive Datalog articulations for which there are no comparable articulations in relational math-algebra and analytics.

CHAPTER 4

SQL

SQL stands for 'Structured Query Language'. It was initially produced for the relational DBMS System R, made by the IBM laboratory facility at 'San Jose' in California amid the late seventies. SQL has since been embraced by numerous different systems; it has been institutionalized and has turn into the reference language for social databases.

SQL is not just a query language. It contains the two elements: a Data Definition Language (DDL) that is responsible for the meaning of a relational database construction and a Data Manipulation Language (DML) that is used to make change and queries on a database. In this part, we will present SQL as the definition language of a database pattern (Segment 4.1). In this regard, we depict the determination of inquiries (Section 4.2) and updates (Section 4.3). In Section 4.4, we portray some more progressed elements of information definition, which require learning of the query language.

In Section 4.5, we present the responsibilities on approval control and we finish with a depiction of the cooperation between SQL and customary programming language (Section 4.6). Some propelled components of DBMSs, which appeared in the previous chapters are moreover upheld by SQL. We will concede the introduction of these parts of the language until the point that they can be presented alongside the suitable ideas.

Hence, we depict the SQL orders for the help of exchanges and the meaning of indices in Chapter 9. Utilization of SQL is to a great extent due to the tremendous measure of institutionalization work that has been dedicated to it, completed principally inside American National Standards Institute (ANSI) and International Organization for Standardization (ISO). Numerous sellers of social frameworks have possessed the capacity to partake in the basic leadership process, as opposed to one merchant specifically having an overwhelming impact. This institutionalization work started in the mid-eighties and is still proceeding today.

Therefore, different renditions of the languages have developed in which the new one is better than the old one. The main definition of a standard for SQL was declared in 1986 by the ANSI. This first standard effectively contained a number of the basic query definition, in the meantime offering a few (constrained) bolster for diagram definition and control. The standard was then reached out in 1989; the most noteworthy expansion to this adaptation was the meaning of referential integrity. This variant is known as SQL - 89.

A second adaptation, which was perfectly compatible with the former one and with the only difference of containing an extensive number of new elements, was distributed in 1992, known as SQL - 92 or SQL - 2 (we will utilize the name SQL – 2). Another variant of the standard, SQL - 3, has as of late been arranged and is called SQL - 99. SQL - 3 is totally compatible with SQL – 2. However, it is still a long way from being broadly embraced. In this consideration, we will dependably allude to SQL - 2 in this book, featuring the new elements that are absent in the prior variant. SQL - 3 incorporates new capacities of late research (among which include dynamic principles and triggers, recursive operations, total operations, new sorts of information, and object paradigm support). Some of these viewpoints are shown in the last piece of the book, which is on managing propelled parts of databases.

Indeed, even without the new SQL - 3 upgrades, SQL - 2 is a rich and complex language, to such an extent that a few years after the presence of the definition report, no business system yet makes accessible all the language highlights. To measure the exact level of consistence with the standard, three levels of many-sided quality of the language developments are characterized. These are referred to separately as Entry-SQL, Intermediate-SQL, and Full-SQL; the systems can be along these lines described by the level that they bolster. Passage SQL is like SQL - 89. It varies just in a couple of slight amendments that were presented amid the meaning of SQL - 2. Middle of the road SQL contains attributes that react best to showcase prerequisites and is offered by a number of the current renditions of relational items. Full SQL contains propelled features that are logically being added to the system.

Then again, the frameworks oftentimes offer components that are definitely not institutionalized. For instance, dynamic standards or triggers are available in a few relational systems yet not in SQL - 2. In any case, the database merchants have picked distinctive grammar and have given diverse semantic understandings to similar elements. This is an issue for two reasons. Right off the bat, the need to pick reflectively one from among the numerous arrangements as of now being executed, urges the alteration of frameworks as of now available and the revamping of utilization produced for them (Chamberlin D. D., 1976).

Also, the presence of more than one proposition for the usage of similar components is a genuine snag to the institutionalization procedure. Indeed, the meaning of gauges is an aggregate procedure that requires all the included gatherings (merchants and in some cases delegates of clients) to reach understanding.

On the off chance that we take a gander at relational systems, we can see that each of them offers an alternate SQL. The distinctions develop most drastically when we analyze their late launched features. In respect to the more united parts of the language, there is a solid adherence to the standard. This enables clients to interface in standard SQL with frameworks that are totally unique in relation to each other, extending from a solitary client DBMS running on the PC up to the actual DBMS on a centralized server putting away the database of a substantial association.

A further imperative perception is that in depicting SQL, we accept that the client associates specifically with the SQL motor keeping in mind the end goal to characterize, refresh, and query the database. With expanding recurrence, system offers interfaces that are anything but difficult to utilize and contain particular projects for the meaning of patterns, updates, and queries. These projects utilize menus and graphical interfaces to produce the relating SQL directions and instructions. In any case, this does not lessen the significance of a learning of the most widely used language of database whole system. This aptitude is vital for the improvement of all non-minor database applications, paying little mind to the capacity of the DBMS interface.

THE DEFINITION IN SQL

During this segment, we present the utilization of SQL for the meaning of database blueprints. Prior to that, we have to represent the documentation we mean to use for the punctuation of the language. Following normal traditions, we will utilize some exceptional symbols.

- Precise sections, ⟨and⟩ are utilized to encase terms.
- Square sections, [and] show that the encased term is discretionary, that is, it may not show up or seem just once.
- Character[varying][(*Length*)][character set *CharSetName*]
- Wavy sections, {and} show that the encased term may not show up or might be rehashed a subjective number of times.
- Vertical bars demonstrate that one among the terms isolated by the bars must show up.

Bended sections (and) should dependably be taken as SQL catchphrases and not as linguistic use definition images.

Elementary Domains

SQL gives six (6) groups of basic areas, which can be utilized to characterize the domain related to the properties of the mapping.

Character

The area character permits the portrayal of a single character or strings. The actual length of characters can be fix or variable. For strings of variable length, the most extreme length is demonstrated. A default character set is determined for every outline (e.g., Latin, Cyrillic, Greek, and Kanji). When it is important to utilize more than one character set, we can indicate it specifically for every area. The sentence structure is:

To characterize an area 'string of 20 characters' with this linguistic structure, we can compose character (20), while an area 'string of Greek letters of variable length, most extreme length 1000', would be signified as character varying (1000) character set Greek. In the event that the length is not determined, the area speaks to a single character. A differing string must determine its most extreme length. Moreover, SQL permits the minimal structures scorch and var-char, for character and shifting character individually.

Bit

This area, presented in SQL - 2, is utilized by characteristics that can expect just the esteem (value) 0 or the esteem 1. The area bit is regularly used to speak to properties, known as banners, which indicate whether a query does or does not have a certain property. Likewise, SQL permits an area 'series of bits' for which the length is indicated as a parameter. At the point when no length is determined, the length of this string is actually set to one. The bit strings can be utilized for the brief portrayal of gatherings of properties. For bits, we can likewise characterize strings of variable length. The sentence structure is:

bit [varying][(*Length*)]

To characterize an area 'string of five bits' or 'series of bits of variable length and greatest length of 100' we can utilize the definitions bit (5) and bit varying (100). The last can be abbreviated to var-bit(100).

Correct numeric areas - This family contains the spaces that permit the portrayal of correct esteems (values) numbers or with a fragmentary part, (for example, average money esteems). SQL makes accessible four distinctive correct numeric spaces:

- The first is : numeric[(Precision [,Scale])]
- The second: decimal[(Precision [,Scale])]
- whole number - (integer)
- Last: smallint

The numeric and decimal spaces speak to numbers with the decimal base.

The parameter 'Precision' determines the quantity of noteworthy digits; utilizing a space decimal (4) we can speak to values among −9,999 and +9,999. Utilizing the parameter 'Scale' we can determine the size of portrayal, that is, we can demonstrate what number of digits ought to show up after the decimal point. In case that we need to incorporate 2 decimal digits, we allot the esteem 2 to Scale. To determine the scale, it is important to indicate the exactness as characterized above. Subsequently, with an area numeric (6,3) we speak to values within −999.999 and +999.999. The distinction between the spaces numeric and decimal lies in the way that the numeric area has precisely the accuracy as shown, while the accuracy of the decimal space ought to be taken as a least prerequisite. Should the accuracy not be determined, the system utilizes default execution esteem. In case that the scale is not determined, it is accepted to be zero.

At the point when the portrayal of parts is not required and an exact control of the extent of the decimal portrayal is not imperative, it ends up noticeably conceivable to utilize the predefined spaces number and smallint. The level of exactness of these spaces is not indicated in the standard rather left to the usage.

Estimated Numeric Spaces

To speak of surmised genuine esteems (helpful, for instance, for speaking to physical amounts), SQL gives the following spaces:

- float [(Precision)]
- twofold exactness
- real

Every one of these domain areas permit the portrayal of genuine numbers by methods for a skimming point portrayal, in which each number compares to a couple of values: the mantissa and the example. The mantissa is a partial esteem, while the example is a whole number. The inexact estimation of the genuine number is acquired by increasing the mantissa by the energy of 10 determined by the example. For instance, the documentation 0.17E16 speaks to the esteem 1.7×10^{15}; also, 0.4E-6 speaks to 4×10^{-7}. A given accuracy can be indicated for the space skim, which speaks to the quantity of digits committed to the portrayal of the mantissa, while the exactness of the example depends on the execution. The area twofold accuracy speaks to the numbers with a more prominent accuracy than the area genuine (Chamberlin D. D., 1974).

Date and Time

This group of spaces and the following were presented in SQL - 2 so as to offer particular help to the administration of transient data, which is critical in numerous applications. They speak to moments of time and involve three (3) structures:

- One: date
- Two: time [(Precision)] [with time zone]

- Three: timestamp [(Precision)] [with time zone]

Each one of these spaces can be organized in fields. The area date permits the field's years, months, and days; the space time permits the field's hours, minutes, and seconds; while, timestamp permits every one of the fields, from year to 2nd. For both, the time and the timestamp we can indicate the exactness that speaks to the number of the decimal places that should be utilized as a part of the portrayal of divisions of a moment. In cas that the exactness is not indicated, time expects an accuracy of zero (determination to the second) and timestamp of 6 (transient determination to the microsecond). On the off chance that the choice with time zone is determined, it moves toward becoming conceivable to get to two fields, 'timezone_hour' and 'timezone_minute'. They speak to the distinction between neighbourhood time and Universal Coordinated Time, earlier known as Greenwich Mean Time. Along these lines 21:03:04+1:00 and 20:03:04+0:00 relate to a similar moment in time, yet the first speaks to it in Middle European Time (varying from the actual base time zone by only +1:00), while the second in Universal-Coordinated -Time.

Fleeting Interims

This group of areas permits the portrayal of interims of time, such as the length of an activity. The language structure is:

interval *FirstUnitOfTime* [to *LastUnitOfTime*]

FirstUnitOfTime and LastUnitOfTime characterize the units of estimation that must be utilized, from the best to the littlest. In this way, we can characterize domain areas, for example, interim year to month to show that the length of the time interim must be measured by the quantity of years and the quantity of months. It must be noticed that the gathering of units of estimation is isolated into two particular gatherings: years and months on one part, and the units from days to seconds on the other part. This detachment happens in light of the fact that it is difficult to think about days and months precisely (given that one month can have days between 28 to 31 days), making it infeasible to look at interims of the two gatherings.

The principal unit that shows up in the definition, whatever it might be, can be portrayed by the exactness that speaks to the quantity of decimal digits utilized as a part of the portrayal. At the point when the littlest unit is the second, we can determine an exactness that speaks to the quantity of decimal spots to be utilized. In the event that the second is the first (and thusly the main) unit, at that point the first parameter speaks to the quantity of noteworthy decimal spots and the second parameter would speak to the quantity of decimal spots of the partial part. At the point when the accuracy is not indicated, it expects the default esteem 2. Along these lines, interim year(5) to month permits the portrayal of interims up to 99,999 years and 11 whole months, while interim day(4) to second(6) permits the portrayal of interims up to 9,999 days, 23hours 59minutes and 59.9999seconds, with an accuracy to a millionth of a moment.

Schema Definition

SQL makes it conceivable to characterize a database outline as a gathering of objects; every outline comprises of an arrangement of spaces, tables, 2 lists, statements, views, and benefits characterized by the accompanying punctuation:

create schema [*SchemaName*][[authorization] *Authorization*]{ *SchemaElementDefinition*}

Approval speaks to the name of the client who possesses the blueprint. On the off chance that the term is discarded, it is expected that the client who issued the charge is the proprietor. The name of the pattern can be precluded and for this situation, the name of the proprietor is embraced as the name of the outline. After the "create schema" order, the client can characterize the outline segments. It is a bit much for every one of the parts to be characterized in the meantime as the mapping is made. This can happen in a few progressive stages. Give us now a chance to see the tables and domain, deferring alternate components of the pattern (statements, views, and benefits) until Chapter 4.4.

The Table Definition

The SQL table comprises of a requested arrangement of properties and of a (conceivably unfilled) set of requirements. For instance, the outline of a table DEPARTMENT is characterized by methods for the accompanying SQL explanation:

create table Department

(

```
        Name        char(20) primary key,
        Address     char(50),
        City        char(20)
)
```

The above table has three characteristics of character string area with the characteristic Name constitutes the essential key of the table.

The sentence structure for the meaning of a table is:

create table TableName

(*AttributeName Domain*[*Default Value*][*Constarints*]

{, *AttributeName Domain*[*Default Value*][*Constarints*]}

[*OtherConstraints*]

)

Each table is characterized by giving its name and the meaning of its traits. Thus, each property has a name and domain and conceivably an arrangement of limitations, which must be fulfilled by the property estimations. When all the qualities have been characterized, we would then be able to characterize the requirements that include more than one quality of the table. A table is at first void and the maker holds every one of the benefits in regards to the table, that is, the rights to get to and to adjust the information.

User Characterized Areas

In the meaning of tables, next to the predefined spaces we have shown in Chapter 4.1.1, it is conceivable to allude to domains that are expressly characterized by the client. Note that there is a cosy connection between the meaning of the spaces of traits and the meaning of the sorts of factors in a programming advanced language (c, Pascal, and so on.). In the two cases, the objective is the meaning of the lawful esteems for information. In any case, there are likewise vital contrasts. From one perspective, the sort constructors in SQL are significantly more restricted than those in programming languages. Then again, be that as it may, SQL offers domain requirements that have no partner in such languages. In SQL, new spaces are determined utilizing the predefined areas portrayed in Segment 4.1.1 by methods for the make of domain command.

Create domain *DomainName* as *ElementaryDomain*

[*DefaultValue*]

[*Constraints*]

A domain is in this manner portrayed by its own particular name, by a basic domain (which can be predefined or already client characterized), by a conceivable default esteem value, and by a (conceivably vacant) arrangement of imperatives that speak to the conditions that must be fulfilled by legitimate space values.

Dissimilar to the sort definition systems of the different programming languages, SQL - 2 does not give area constructors, for example, record or cluster (other than the likelihood of characterizing series of characters or bits). This limitation gets from the relational information model, which requires that all traits have rudimentary spaces.

The assertion of areas relates a domain name with an arrangement of imperatives. This winds up noticeably essential when, for instance, we have to rehash a similar characteristic definition in a few tables. The meaning of a domain empowers its reuse and makes trait definitions more effectively modifiable. We can change only the area definition (especially the default and the imperatives related with it) and such a change is engendered to all of the tables where it is really utilized.

Default Values

In the language structure for characterizing domains and tables, we take note the nearness of a term DefaultValue, related with areas and qualities. This term determines the values that the quality must expect when a line is embedded into the table without an esteem being determined for the property itself. At the point when default esteem is not determined, the esteem invalid is accepted as default.

The linguistic structure for the particular default esteems is:

default <*Generic*Value|user|null>

GenericValue - speaks to an esteem that is perfectly compatible with the domain, given as a constant or characterized as the consequence of the assessment of an articulation. The alternative client sets as default esteem the login name of the client who issues the order to refresh the table. Whenever a characteristic has a domain with default value and is unequivocally given an alternate default esteem, the default esteem related with the characteristic wins and turns into the successful default esteem.

For instance, a trait NumberOfChildren, which permits a whole number as an esteem and which has the default esteem zero is characterized by:

NumberOfChildren smallint default 0

In the event that, amid the inclusion of a line, the estimation of the trait is not indicated, it is given to be zero.

Intra-Relational Constraints

In the meaning of the two, domain and table, we can characterize limitations, like, properties that must be checked by each occasion of the database. We presented the constraints in Section 2.2, making the qualification between intra-relational constraints (which include a solitary connection) and between relational limitations (which consider a few relations). The most effective develop for the particular of non-specific imperatives, both between relational what's more, intra-relational, is that of check, which, in any case, requires the particular of questions-queries to the database. We talk about this later, in Segment 4.4, once we have represented SQL queries-questions. In this segment, we will outline predefined intra-relational imperatives.

The most straightforward intra-relational imperatives are not null, unique, and primary key.

Not Null

The null value is a unique value, which shows the nonattendance of data. A null value can for the most part speak to various circumstances as we talked about in Section 2.1.5.

SQL - 2 does not permit the qualification among the different translations of the null value. Those applications that should have the capacity to recognize among these different circumstances must depend on impromptu arrangements, for example, the presentation of different properties.

The constraint that is not null shows that the null value is not acceptable as the property value. For this situation, the property should dependably be indicated for the most part at the addition arrange. In any case, assuming default value other than null is related with the trait, it ends up plainly conceivable to complete an inclusion even without giving an incentive to the property since the default value will be allotted to it consequently.

The constraint is indicated by adding to the characteristic definition the catchphrases not null:

Surname character (20) not null

Unique

A unique constraint is connected to a quality (or an arrangement of properties) of a table and forces the necessity that the property (or the arrangement of traits) is a super-key. Along these lines, the requirement guarantees that diverse columns do not have similar values. An exemption is made for the null value, which can show up in different columns without damaging the requirement, since it is expected that every null value speaks to an obscure real value unique from that of another null value.

This constraint can be characterized in two ways. The main option can be utilized just when the imperative includes a solitary property. For this situation, the detail of the trait is trailed by the keyword unique (like the detail of the not null constraint):

RegistrationNumber character(6) unique

The second option is essential when we have to characterize the constraint on an arrangement of characteristics. For this situation, after the meaning of the characteristics, we utilize the condition:

unique (*Attribute*{ *Attribute* })

Below is a case of the utilization of this condition:

FirstName character(20) not null,

Surname, character(20) not null,

unique (Surname, FirstName)

It ought to be noticed that the above definition is altogether different from a definition. Below is an example.

FirstName character(20) not null unique,

Surname, character(20) not null unique

In the principal case, the imperative forces the condition that there can be no two rows that do have both a similar first name and a similar surname. In the second (stricter) case, the requirement is damaged if either a similar first name or a similar surname is shown more than one time.

Primary Key

As we talked in Section 2.2, it is generally critical to determine a primary key, the most vital identifier for a connection.

Likewise, SQL enables a primary key constraint to be indicated once for each table (while it is conceivable to utilize the requirements one of a kind and not null a discretionary number of times). Like the one of a kind limitation, the primary key constraint can be straightforwardly characterized on a solitary property or it can be characterized by posting the few traits that make up the primary key. None of the traits of a primary key can expect the null value. Along these lines, the meaning of primary key infers a certain definition not null for every one of the characteristics of the primary key (Cannan, 1992).

For instance, the accompanying definition forces the imperative that the combination of properties FirstName and Surname constitute the primary key:

FirstName character(20),

Surname character(20),

Dept character(15),

Salary numeric(9) default 0,

Primary key (Surname, FirstName)

Inter-Relational Constraints

As we found in Section 2.2, the most essential between relational requirements are referential integrity requirements. In SQL, the proper build to characterize them is the remote key requirement.

This constraint makes a connection between the estimations of the attributes of a table and the estimations of the attributes of another table. As for such a limitation, we call the included tables inner and outside. The limitation is that for each line of the inner table the estimation of a given property, if not the same as the null value, must be available among the values of guaranteed characteristic of the lines having a place with the outer table. The main prerequisite that the sentence structure forces is that ascribe alluded to in the outer table is liable to a one of a kind imperative, that is distinguishes the tuples of the outside table. This trait speaks to the primary key of the actual table for which the one of a kind imperative is ensured. A few qualities might be engaged with the imperative, when the key for the outer table comprises of a set of qualities. For this situation, the main distinction is that it is important to think about tuples of qualities as opposed to single values.

The requirement can be characterized in two ways similar to the unique and primary key constraints. On the off chance that there is just a single property included, it is conceivable to characterize it utilizing the syntactic develop references, which shows the outside table and property. The more broad definition, which is fundamental when the connection is spoken to by an arrangement of qualities, utilizes the developed remote key linguistically put after characteristic definitions. This develops the compelled properties of the inward table engaged with the connection trailed by the name of the outside table and names of the referenced properties. Let us give a case of the principal use.

```
create table Employee
(
RegNo       character(6) primary key,
FirstName   character(20) not null,
Surname     character(20) not null,
Dept        character(15)
            References Department(DeptName),
Salary      numeric(9) default 0,
City        character(15),
unique  (Surname, FirstName)
)
```

The constraint indicates that the quality Dept can accept just a single of the values that the columns of the DEPARTMENT table have for the DeptName characteristic.

On the off chance that we require the ascribes FirstName and Surname to show up in a table of individual records, we have to utilize the second option, which is:

```
create table Employee
(
RegNo        character(6) primary key,
FirstName    character(20) not null,
Surname      character(20) not null,
Dept         character(15)
             References Department(DeptName),
Salary       numeric(9) default 0,
City         character(15),
unique  (Surname, FirstName),
foreign key(FirstName, Surname)
        references PersonalRecord(FirstName, Surname)
)
```

The correspondence between the local and external qualities reflects their request: the primary trait in the remote key compares to the first quality in the referenced table, et cetera for alternate properties. In this case, FirstName and Surname of EMPLOYEE relate separately to FirstName and Surnames of PERSONALRECORD.

On account of the considerable number of imperatives seen up until this point, when the framework identifies infringement created by a refresh, the framework just rejects the update, flagging the blunder to the client. With referential respectability requirements, SQL additionally enables the client to pick different moves to be made when an infringement is presented.

We outline the point by a case. Consider the meaning of the remote key limitation on the trait Dept in the table EMPLOYEE. The limitation can be disregarded by working either on the columns of the interior table

EMPLOYEES or on those of the outside table DEPARTMENTS. There are, as it were, two approaches to present infringement by changing the substance of the inner table: by embedding another line or by changing the estimation of the alluding trait. No specific help is offered on account of both of these infringements; the operation will basically be rejected.

Then again, different alternatives are offered for reacting to infringement produced by changes to the outside table. The purpose behind this asymmetry is because of the specific significance of the outside table, which from the application perspective ordinarily speaks to the key table (or ace-master).

The interior table (or slave) must adjust to varieties in the ace. In certainty, all activities will create an intercession on the inside table. The operations on the outer table that can create infringement are the refresh of estimations of the referenced characteristics and the cancellation of lines (cancellation of lines in DEPARTMENTS and refresh of the quality DeptName). The sort of response can vary as per the summon that created the infringement.

Specifically, for updates it is conceivable to respond in one of the accompanying ways.

- Course-cascade—the new estimation of the characteristic of the outside table is allocated to all the coordinating lines of the interior table.
- Set-invalid—the esteem invalid is doled out to the alluding trait in the inner table set up of the esteem changed in the outside table.
- Set-default—t he actual default esteem is doled out to the alluding trait in the interior table set up of the esteem adjusted in the outer table.
- No activity—the refresh is essentially dismissed, with no amending activity by the system.

For infringement delivered by the cancellation of a component of the outside table, a similar arrangement of responses is accessible.

- Course-cascade—every one of the columns of the inner table comparing to the erased push are likewise erased.
- Set-invalid—the esteem invalid is relegated to the alluding

quality set up of the esteem erased from the outer table.

- Set-default—the actual default esteem is relegated to the alluding quality in place of the esteem erased from the outer table.
- No activity—the erasure is rejected.

We can indicate distinctive approaches for various occasions (for instance, utilizing a course approach for refreshes and a set invalid arrangement for erasures).

The utilization of the course approach accepts that the rows of the inward table are firmly connected to the comparing lines of the outer table. Thus, on the off chance that adjustments are made to the outside table, similar alterations must be made to every one of the lines of the interior table. Then again, the other strategies expect a weaker reliance between the columns of the two tables.

The response strategy is indicated quickly after the respectability limitation, as indicated by the syntax below.

```
on <delete | update>
        <cascade |set null| set default| no action>
```

The case below is a referential integrity limitation in which the repair policies are actually left null for erase and course for updates:

```
create table Employee
(
        RegNo         character(6),
        FirstName     character(20) not null,
        Surname       character(20) not null,
        Dept          character(15)
        Salary        numeric(9) default 0,
        City          character(15),
        Primary key(RegNo),
        foreign key(Dept) references Department(DeptName)
```

on delete set null

on update cascade,

unique (Surname, FirstName)

)

Schema Refreshes

SQL gives primitives to the control of database mappings, which empower the alteration of table definitions. The alter command includes add and drop.

The add Command

This permits the alteration of areas and patterns of tables. Below are different statements to initiate this command.

alter domain *DomainName* <set default *DefaultValue*>

drop default |

add constraint *ConstarintDef* |

drop constraint *ConstarintName*

alter table *TableName*

alter column AttributeName

<set default *DefaultValue* | drop default>|

add constraint *ConstraintDef* |

drop constrain *Constraint* |

add column *AttributeDef* |

drop column *AttributeName* >

By initiating alter domain and alter table, we can add and drop limitations as well as change default esteems related with areas and traits. Besides, we can add and drop traits and limitations inside the diagram of a table. Note that when another requirement is characterized, it must be fulfilled by the information officially displayed in the database. In the event that the database contains infringement of the new limitation, the requirement definition will be rejected.

For instance, the accompanying summon expands the pattern of the table Division with a trait NoOfOffices that makes it conceivable to speak to the quantity of workplaces inside the division, as shown below.

alter table Department add column NoOfOffices numeric(4)

The drop Command

While the add command adds areas or compositions of the tables, the drop command permits the evacuation of parts regardless whether they be patterns, domains, tables, views, or affirmations. Declarations are imperatives that are not related with any specific table. These will be introduced in Section 4.4. The order has the punctuation as presented below.

drop <schema | domain | table | view | assertion> ComponentName

[restrict | cascade]

The limit alternative determines that summon must not be completed if the part being erased is not vacant. Subsequently, a diagram is not expelled if it contains tables or different components. A space is not expelled in case that it shows up in the table definition. The table is not expelled in the event that it has rows or in case that it is available in a meaning of a table or view. Also, a view is not expelled on the off chance that it is utilized as a part of the meaning of different perspectives. The confine choice is the default.

With the course choice, the segment is evacuated together with the segments relying upon it. Therefore, when a non-exhaust outline is evacuated, every one of the objects of which it is built is likewise dispensed with. By evacuating a space that shows up in the meaning of a property, the course choice makes the name of the space be evacuated. However, the properties that were characterized utilizing that space remain related with a similar essential area definition. Consider for instance the area LongString characterized as char (100). On the off chance that LongString is wiped out (by methods for the charge drop space LongString course) every one of the properties characterized on that space will specifically expect the space char (100). At the point when a table is expelled with the course choice, every one of its lines is lost. In the event that the table showed up in another meaning of a table or view, this would likewise be expelled. By dispensing with a view that shows up in different tables or perspectives, these too are evacuated.

The course choice, more often than not, creates a chain response. Every one of the components contingent upon a component that is wiped out are wiped out in their turn, until there are no outstanding components that contain in their definitions components that have been dispensed with. It is important to practice extraordinary alert in the utilization of this choice since it is conceivable that inferable from some disregarded reliance. The charge could have an alternate outcome from the one expected. Numerous systems make it conceivable to test the consequence of the drop cascade command before it is really executed (Date C. J., 1993).

Relational Indexes

Albeit just not completely indicated by the standard, each relational DBMS deals with its own information lexicon (or rather the depiction of the tables exhibit in the database) utilizing a relational diagram. The database consequently contains two sorts of table: those that contain the information and those that contain the metadata. This second gathering of tables constitutes the inventory of the database.

This normal on relational system usage is known as 'reflexivity'. A DBMS commonly deals with the inventory by utilizing structures like those in which the database case is put away. Consequently, an object-oriented database for instance will have an information word reference that is characterized on a query. Along these lines, the database can utilize the same capacities for the inside association of metadata as utilized for the administration of the database occurrence.

The definition and alteration summons of the database pattern could, in principle be supplanted by control summons that work straightforwardly on the tables of the information lexicon, making unnecessary the presentation of unique commands for the definitions and meanings of the pattern. This is not done for two reasons. The first reason is the nonattendance of an institutionalization of the word reference, which varies enormously starting with one item then onto the next. The second reason is the need of guaranteeing that the summons for the control of constructions is clear and quickly unmistakable and linguistically discernible from the charges that change the database occasion.

The SQL - 2 standards for the information word reference depends on a two-tier portrayal. The primary level is the 'DEFINITION_SCHEMA', which is made up of a gathering of tables that contain the portrayals of the considerable number of structures in the database. The gathering of tables showing up in the standard is definitely not utilized by any execution, since the tables give a portrayal of as it were those parts of a database secured by SQL - 2. What is forgotten is all the data concerning the capacity structures, which regardless of the fact that not are introduced in the standard, frame a central piece of a blueprint. The tables of the standard, in this manner frame a layout to which the systems are prompted (not obliged) to acclimate. The second segment of the standard is the 'INFORMATION_SCHEMA'. This comprises of a gathering of perspectives on the 'DEFINITION_SCHEMA'. These perspectives completely constitute some portion of the standard and frame an interface for the information word reference, which should be offered by the systems that need to be fully compatible with this standard. The 'INFORMATION_SCHEMA' contains perspectives, for example, DOMAIN, 'DOMAIN_CONSTRAINTS', TABLE, VIEW, and COLUMN, up to a sum of 23 views that depict the structure of the database.

Instead of depicting the structure of these tables, we give a disentangled case of the substance of one of these perspectives. In the Figure 51, we can actually see the streamlined substance of the COLUMNS perspective of the inventory depicting the tables EMPLOYEE and DEPARTMENT.

Table_Name	Column_Name	Ordinal_Position	Column_Default	Is_Nullable
Employee	RegNo	1	NULL	N
Employee	FirstName	2	NULL	N
Employee	Surname	3	NULL	N
Employee	Dept	4	NULL	Y
Employee	Salary	5	0	Y
Employee	City	6	NULL	Y
Department	DeptName	1	NULL	N
Department	Address	2	NULL	Y
Department	City	3	NULL	Y

Figure 51: The View for the COLUMNS.

(Source: Melton, J. S. Understanding the New SQL.)

In Figure 52, we see a case of the reflexivity of the information word reference with the depiction in COLUMN of the view itself.

Table_Name	Column_Name	Ordinal_Position	Column_Default	Is_Nullable
Columns	Table_Name	1	NULL	N
Columns	Column_Name	2	NULL	N
Columns	Ordinal_Position	3	NULL	N
Columns	Column_Default	4	NULL	Y
Columns	Is_Nullable	5	Y	N

Figure 52: The Description for the COLUMNS.

(Source: Melton, J. S. Understanding the New SQL.)

SQL QUERIES

The piece of SQL committed to the statement of inquiries is incorporated into the DML. In any case, the partition between the DDL and the DML is not unbending and part of the query definition sentence structure will likewise be utilized as a part of certain propelled components of the construction (see Chapter 4.4).

The Explanatory Idea of SQL

SQL communicates queries predominantly, which is by determining the properties of recovered information and not how to acquire it. In this regard, SQL takes after the standards of social analytics and appears differently in relation to procedural question languages such when relational algebra in which an information recovery system must be indicated in the inquiry. The SQL inquiry is passed for execution to the inquiry streamlining agent. This is a DBMS segment that breaks down the question, chooses an inquiry execution methodology, and figures an equal query in the inward procedural language of the database administration framework. This procedural language is avoided (hidden) from the client. Whoever composes queries in SQL can overlook the interpretation and advancement angles.

The huge exertion that has gone into the advancement of improvement strategies has made very effective inquiry execution feasible for most relational DBMSs.

There are a wide range of approaches to express a query in SQL. The software engineer must settle on decisions that are constructing not with respect to proficiency but on qualities, for example, intelligibility and simplicity of adjustment of the inquiry. Along this, SQL encourages the software engineer's work, making conceivable the depiction of theoretical and abnormal state inquiries.

Simple Queries

Queries operations in the SQL are determined by methods for the select articulation. Let us initially see the basic structure of a select.

The three sections that make up a select direction are frequently called the 'objective list', the 'from condition', and the 'where condition'. A more point by point portrayal of a similar linguistic

select *TargetList*

from *TableList*

[where *Condition*]

structure is as follows.

An SQL query considers the lines that have a place with the Cartesian result of the tables recorded

select *AttrExpr* [[as] *Alias*]{, *AttrExpr* [[as] *Alias*]}

from *Table* [[as] *Alias*] {, *Table* [[as] *Alias*]}

[where *Condition*]

in the from condition, and chooses those that fulfil the conditions communicated in the where statement. The aftereffect of the execution of an SQL inquiry is a table, with a line for each row chosen by the where condition, in which sections result from the assessment of the articulations AttrExpr that show up in the objective list. Every section can be renamed by methods for the Alias that shows up promptly after the articulation. The tables can likewise be renamed utilizing the Alias. Consider a database containing the tables EMPLOYEE (FirstName, Surname, Dept, Office, Salary, City) and DEPARTMENT (DeptName, Address, City).

EMPLOYEE	FirstName	Surname	Dept	Office	Salary	City
	Mary	Brown	Administration	10	45	London
	Charles	White	Production	20	36	Toulouse
	Gus	Green	Administration	20	40	Oxford
	Jackson	Neri	Distribution	16	45	Dover
	Charles	Brown	Planning	14	80	London
	Laurence	Chen	Planning	7	73	Worthing
	Pauline	Bradshaw	Administration	75	40	Brighton
	Alice	Jackson	Production	20	46	Toulouse

Figure 53: Employee Table Content.

(Source: Melton, J. S. Understanding the New SQL.)

Query-1: Find the compensation of the representatives named Brown.

select Salary as Remunration

from Employee

where Surname = 'Brown'

On the off chance that there are no representatives named Brown, the inquiry displays an empty result. Otherwise, it restores a table with the same number of rows as there are representatives with that exact surname. By applying the query, we get the result as seen in Figure 54. There are two representatives with the surname Brown, hence outcome contains two lines.

Remu-neration
45
80

Figure 54: Query 1 Results.

(Source: Melton, J. S. Understanding the New SQL.)

Give us now a chance to proceed with the examination of SQL queries, continuously presenting more unpredictable develops.

Target List

The objective list determines the components of the diagram of the coming about tables. The uncommon character * (asterisk) can likewise show up in the target list, returning all the properties of the tables recorded in the 'from' condition.

Query-2: Find all the records of employees with the surname Brown. The result is shown in Figure 55.

select *

From Employee

Where Surname = 'Brown'

FirstName	Surname	Dept	Office	Salary	City
Mary	Brown	Administration	10	45	London
Charles	Brown	Planning	14	80	London

```
select Salary/12 as MonthlySalary
from Employee
where Surname = 'White'
```

MonthlySalary
3.00

Figure 55: Query 2 Results.

(Source: Melton, J. S. Understanding the New SQL.)

Query-3: Find the monthly compensation of employees with the surname White. The outcome is presented in Figure 56.

select Salary / 12 as MonthlySalary

from Employee

where Surname = 'White'

MonthlySalary
3.00

Figure 56: Query 3 Results.

(Source: Melton, J. S. Understanding the New SQL.)

From Provision

When we have to figure an inquiry that includes lines having a place with more than one table, the contention of the provision is given as a rundown of tables. The conditions in the where provision are connected to the Cartesian result of these tables. A join can be determined by unequivocally demonstrating examinations between properties of various tables.

Query-4: Find the names of employees and their corresponding cities based on the department.

select Employee.FirstName, Employee.Surname, Department.City

from Employee, Department

where Employee.Dept = Department.DeptNmae

DEPARTMENT

DeptName	Address	City
Administration	Bond Street	London
Production	Rue Victor Hugo	Toulouse
Distribution	Pond Road	Brighton
Planning	Bond Street	London
Research	Sunset Street	San Jose

Figure 57: Department Table Content.

(Source: Melton, J. S. Understanding the New SQL.)

Taking the substance of EMPLOYEE and DEPARTMENT separately from Figure 53 and Figure 57, the consequence of the assessment of Query 4 is the table appeared in Figure 58.

In the above inquiry, we take note of the utilization of the speck administrator to recognize the tables from which properties are taken. For instance, Employee.Dept indicates the Dept property of the table EMPLOYEE. This utilization is regular in numerous programming languages, to distinguish the fields of an organized variable. It is important to utilize this documentation when the tables recorded in the 'from' statement have traits with a similar name, keeping in mind the end goal to recognize among the references to the homonym characteristics. At the point when there is no threat of uncertainty, in light of the fact that the trait name shows up in just a single of the tables, we can indicate the characteristic without announcing the table to which it has a place.

FirstName	Surname	City
Mary	Brown	London
Charles	White	Toulouse
Gus	Green	London
Jackson	Neri	Brighton
Charles	Brown	London
Laurence	Chen	London
Pauline	Bradshaw	London
Alice	Jackson	Toulouse

Figure 58: Query 5 Results.

(Source: Melton, J. S. Understanding the New SQL.)

Query-5: The main homonym characteristic in the tables EMPLOYEE and DEPARTMENT is the trait City. The previous question can in this manner be communicated as takes after, by utilizing an assumed name for table DEPARTMENT with the aim of contracting the reference to it:

select FirstName. Surname, D.City

from Employee, Department as D

where Dept = DeptName

'Where' Statement

The condition in the where statement is a boolean articulation built by joining basic predicates with the operator *and, or,* and *not*. Every straightforward predicate utilizes the operators =, < >, <, >, <=, and >= to assemble an examination that has on one side of an articulation shaped from the estimations of the properties of the row and, on the opposite side, a consistent value or another articulation. The syntax offers priority to the operator *not* in the assessment, but not between *and* and *or*. In the event that we have to express an inquiry that requires the utilization of both *and* and *or*, we need to demonstrate the request of assessment by utilizing brackets.

Query-6: Find the first name and surname of employees who work in office number '20' under the Administration department.

select FirstName, Surname

from Employee

where Office = '20' and Dept = 'Administration'

The outcome in Figure 59 is taken from the database in Figure 53.

FirstName	Surname
Gus	Green

Figure 59: Query 6 Results.

(Source: Melton, J. S. Understanding the New SQL.)

Query-7: Find the first name and surname of employees who work either in the Administration department or in the Production department.

select FirstName, Surname

from Employee

where Dept = 'Administration' or

Dept = 'Production'

By applying the query to the table in Figure 53, we get the outcome as shown in Figure 60.

FirstName	Surname
Mary	Brown
Charles	White
Gus	Green
Pauline	Bradshaw
Alice	Jackson

Figure 60: Query 7 Results.

(Source: Melton, J. S. Understanding the New SQL.)

Query-8: Find the primary names of employees with the surname 'Brown' who work in the Administration department or the Production department. The outcome is appeared in Figure 61.

```
select FirstName
from Employee
where Surname = 'Brown' and
        (Dept = 'Administration' or
Dept = 'Production')
```

FirstName
Mary

Figure 61: Query 8 Results.

(Source: Melton, J. S. Understanding the New SQL.)

In addition to the typical predicates for relational examinations, SQL gives an operator like the correlation of strings. This operator performs design coordinating with somewhat determined strings, acquired by utilizing the extraordinary characters _(underscore) and %(percent). The underscore speaks to a subjective character in the correlation, while the percent sign speaks to a string (perhaps vacant) of subjective characters. The correlation like 'ab%ba_' will along these lines be fulfilled by any series of characters.

Query-9: Find the employees with the surname having 'r' as the second letter and ending in 'n'. The outcome is presented in Figure 62.

```
select *
from Employee
where Surname like '_r%n'
```

FirstName	Surname	Dept	Office	Salary	City
Mary	Brown	Administration	10	45	London
Charles	Brown	Planning	14	80	London
Gus	Green	Administration	20	40	Oxford

Figure 62: Query 9 Results.

(Source: Melton, J. S. Understanding the New SQL.)

Administration of Null Values

As we found in Section 2.1.5, a null value in a characteristic can imply that a specific trait is not relevant or the esteem is material but obscure.

Attribute is [not] null

For the determination of terms with invalid esteems, SQL supplies the invalid predicate, the syntax of which is as show below.

Salary > 40

The predicate is null and gives a genuine outcome if the quality has the null value. The 'is not null' predicate creates the inverse outcome. Null values particularly affect the assessment of ordinary predicates. Consider a straightforward predicate for the correlation of the estimation of a characteristic with a consistent value.

This predicate will be valid for the rows in which the compensation property is greater than 40. Remember what we mentioned in Section 3.1.8, there are two distinct answers for managing the circumstance in which the SALARY characteristic has the null value. The primary arrangement, which was embraced by SQL - 89 utilizes conventional two-valued rationale and essentially enables the predicate to be viewed as false. The second arrangement is the one embraced by SQL - 2. This uses a three-valued rationale, in which a predicate restores the obscure value when actually any of the terms of the real predicate has an invalid esteem. Note that the invalid predicate is a special case, since it generally returns both of the qualities true or false, never the obscure esteem.

The contrast between the arrangements in view of a few esteemed rationale rises just when complex articulations are assessed. Now and again, the conduct of the framework when invalid esteems are available can be a long way from instinctive, especially when complex predicates are developed utilizing refutation or settled queries (presented in Section 4.2.6).

Logarithmic Understanding of SQL Inquiries

We can build a correspondence between SQL queries and the proportional inquiries communicated in relational algebra.

Given a query in SQL in its least complex frame:

select $T_1.Attribute_{11}$, … $T_h.Attribute_{hm}$

from $Table_1.T_1$, …, $Table_n$ T_n

where *Condition*

we can utilize the accompanying interpretation, where preparatory renamings (excluded here for effortlessness) are connected to each $TABLE_i$ with the goal that joins are in fact Cartesian items:

$$\pi_{T1 \cdot Attrribute11,\ldots,ThAttributehm}(\sigma_{condition}(TABLE_1 \bowtie \ldots \bowtie TABLE_n))$$

For more intricate SQL inquiries, the transformation equation appeared above is no longer straightforward. We could, in any case, exhibit a system for making an interpretation of a SQL question into a proportional inquiry in relational algebra.

The connection among SQL and tuple analytics with extend presentations is even more grounded (Section 3.2.3).

Expecting that T_1, T_2, … , T_h show up in the objective rundown, while T_{h+1}, T=, … , T_n do not, the nonspecific articulation select implies that is equivalent to that of the accompanying articulations of tuple math with go announcements:

$\{t_1.Attribute_{11}, \ldots, t_h.Attribute_{hm}|$

$t1(TABLE_1), \ldots, t_h(TABLE_h)$

$|\exists t_{h+1}(TABLE_{h+1}), (\ldots,(\exists t_n(TABLE_n)(Condition'))\ldots)\}$

where Condition′ is the recipe got from Condition, substituting the relational algebra math documentation into the SQL one.

It ought to be noticed that these correspondences among SQL and the other languages do not hold any more in cas that we think of some as cutting edge components of SQL such as the assessment of amassed administrators (Section 4.2.3). Consequences of SQL inquiries and of arithmetical or analytics articulations contrast on copy tuples, as discussed below.

Copies

A noteworthy distinction among SQL, relational algebra, and calculus is the administration of copies. In relational algebra and analytics, a table is seen as a connection from the scientific perspective and in this way as an arrangement of components (tuples) unique in relation to each other. In SQL, a table might have many lines with similar esteems for every one of their traits. These are known as copies. For imitating the conduct of social variable based math in SQL, it would be important to wipe out every one of the copies every time a projection operation is delivered. The expulsion of copies is tedious and frequently superfluous, in that much of the time the outcome contains no copies. To clarify, when the outcome incorporates a key for each table that shows up in the from provision, the subsequent table cannot contain indistinguishable rows. Hence, it is chosen to permit the nearness of copies in SQL, abandoning it to the individual composition the question to indicate when the expulsion of copies is essential.

The disposal of copies is indicated by the watchword unmistakable, set instantly after the watchword select. The sentence structure likewise takes into account the particular of the watchword all set up of unmistakable, showing the necessity that the copies ought to be held. The utilization of the word all is discretionary, in that the preservation of copies constitutes the default alternative (Melton J. S., 1993).

Given the table PERSON(TaxCode, FirstNames, Surnames, City) as shown in Figure 63, we wish to retrieve the data of city with the surname Brown. We will present two illustrations, the first permits the nearness of copies while the second uses the unmistakable alternative and expels them.

PERSON

TaxCode	FirstName	Surname	City
BRWMRA55B21T234J	Mary	Brown	Verona
LBLCLR69T30H745Z	Charles	Leblanc	Paris
BRWGNN4IA3IB344C	Giovanni	Brown	Verona
BRWPRT75CI2F205V	Pietro	Brown	Milan

Figure 63: Person Table Content.

(Source: Melton, J. S. Understanding the New SQL.)

Query-10: Find the city of those whose surname is Brown.

select City

from Person

where Surname = 'Brown'

Query-11: Find the city of those whose surname is Brown, showing the name of the city only once.

select distinct City

from Person

where Surname = 'Brown'

By executing the two queries above on the table shown in Figure 63, we get the outcomes as presented in Figure 64.

City
Verona
Verona
Milan

City
Verona
Milan

Figure 64: Query 10 + 11 Result.

(Source: Melton, J. S. Understanding the New SQL.)

Internal and External Joins

An option punctuation presented in SQL - 2 for the determination of joins makes it conceivable to recognize the conditions that speak to join conditions and those that speak to determinations of lines. Along this, we can likewise determine external joins and different expansions.

By using this kind of syntax, the join conditions do not show up as the contention of the 'where' clause, rather is moved into the 'from' statement related with the tables that are associated with the join.

select *AttrExpr* [[as] *Alias*] {, *AttrExpr*[[as] *Alias*]}

from *Table* [[as] *Alias*]

{[*JoinType*] join *Table* [[as] *Alias*] on *JoinCondition*}

[where *OtherCondition*]

The parameter JoinType indicates which kind of join to utilize. For this reason, we then can substitute the terms inward, right-external, left-external, or full-external (the term external is discretionary). The internal join is the customary theta-join of relational algebra.

Query-12: Query-5 can be reworked utilizing the language structure of the inward join in the following way.

select FirstName, Surname, D.City

from Employee inner join Department as D on Dept = DeptName

With the internal join between the two tables, the rows associated with the join are a subset of the lines of each table. It can happen that few lines are excluded on the grounds that no row in the other table exists for which the condition is met. This property frequently clashes with the requests of uses that may need to hold the lines that would be dispensed by the join. In composing the application, we may like to utilize invalid qualities to speak to the nonappearance of data in the other table. As we have found in Section 3.1.5, the external join has accurately executing a join while keeping up every one of the lines that show up in one or both the tables included.

There are three distinct sorts of external join: left, right, and full. The left join gives an indistinguishable outcome from the inward join. However, the former incorporates the lines of the table that shows up in the left part of the join for which no comparing lines exist in the right-hand table. The correct join carries on symmetrically (keeps the lines of the right-hand table). While, the full join gives the consequence of the internal join alongside the columns barred from the two tables. Consider the tables DRIVER and AUTOMOBILE appeared in Figure 65.

Query-13: Find the drivers with and without automobile.

select FirstName, Surname, Driver.DriverID, CarRegNo, Make, Model from Driver left join Automobile on

(Driver.DriverID = Automobile.DriverID)

The outcome is appeared in Figure 66. Note the last line that record shows a driver without data about automobile.

Query-14: Find every driver and every automobile, demonstrating the conceivable connection between them.

DRIVER

FirstName	Surname	DriverID
Mary	Brown	VR 2030020Y
Charles	White	PZ 1012436B
Marco	Neri	AP 4544442R

AUTOMOBILE

CarRegNo	Make	Model	DriverID
ABC 123	BMW	323	VR 2030020Y
DEF 456	BMW	Z3	VR 2030020Y
GHI 789	Lancia	Delta	PZ 1012436B
BBB 421	BMW	316	MI 2020030U

Firstname	Surname	DriverID	CarRegNo	Make	Model
Mary	Brown	VR 2030020Y	ABC 123	BMW	323
Mary	Brown	VR 2030020Y	DEF 456	BMW	Z3
Charles	White	PZ 1012436B	GHI 789	Lancia	Delta
Marco	Neri	AP 4544442R	NULL	NULL	NULL

Figure 65: Tables of Driver and Automobile.

(Source: Melton, J. S. Understanding the New SQL.)

DRIVER

FirstName	Surname	DriverID
Mary	Brown	VR 2030020Y
Charles	White	PZ 1012436B
Marco	Neri	AP 4544442R

AUTOMOBILE

CarRegNo	Make	Model	DriverID
ABC 123	BMW	323	VR 2030020Y
DEF 456	BMW	Z3	VR 2030020Y
GHI 789	Lancia	Delta	PZ 1012436B
BBB 421	BMW	316	MI 2020030U

Firstname	Surname	DriverID	CarRegNo	Make	Model
Mary	Brown	VR 2030020Y	ABC 123	BMW	323
Mary	Brown	VR 2030020Y	DEF 456	BMW	Z3
Charles	White	PZ 1012436B	GHI 789	Lancia	Delta
Marco	Neri	AP 4544442R	NULL	NULL	NULL

Figure 66: Query 16 Results.

(Source: Melton, J. S. Understanding the New SQL.)

select FirstName, Surname, Driver.DriverID, CarRegNo, Make, Model
from Driver full join Automobile on
(Driver.DriverID = Automobile.DriverID)

The query results to the table appeared in Figure 67. Note the last line of the table depicts a vehicle for which there is no data about the driver.

Firstname	Surname	DriverID	CarRegNo	Make	Model
Mary	Brown	VR 2030020Y	ABC 123	BMW	323
Mary	Brown	VR 2030020Y	DEF 456	BMW	Z3
Charles	White	PZ 1012436B	GHI 789	Lancia	Delta
Marco	Neri	AP 4544442R	NULL	NULL	NULL
NULL	NULL	NULL	BBB 421	BMW	316

Figure 67: Query 14 Results.

(Source: Melton, J. S. Understanding the New SQL.)

A few usage of SQL indicate the external join in a non-standard manner by including a specific character or arrangement of characters (for instance * or (+)) to the traits engaged with the join condition. Thus, we can figure the external join without utilizing the sentence structure above. For instance, Query-13 could be figured as presented below.

Query-15:

select FirstName, Surname, Driver.DriverID, CarRegNo, Make, Model
from Driver, Automobile
where Driver.DriverID * = Automobile.DriverID

These arrangements being outside the SQL - 2 standard are not compact from one framework to another.

A further element of SQL - 2 is the likelihood of utilizing the catchphrase characteristic prior to the join sort. This makes it conceivable to characterize the normal join of relational algebra. In this way, in the joining of two tables, we might utilize a verifiable state of fairness on every one of the traits having a similar name (see Area 3.1.5). For instance, Query-14 could be composed as found below.

Query-16:

select FirstName, Surname, Driver.DriverID, CarRegNo, Make, Model from Driver natural full join Automobile

Notwithstanding the upside of an expanded conservativeness, the characteristic join is not ordinarily accessible on business systems. One actual reason is that a real query that uses a characteristic join can acquaint dangers with the applications, since its conduct can change essentially because of little minor departure from the blueprint. Another reason is that the common join makes it important to examine totally the blueprint of the tables engaged with request to comprehend the join condition. This is a detriment when composing and when perusing the inquiry, since in the two circumstances it is important to do a cautious correlation of the diagrams of the joined tables making sure of the conduct of the query.

Utilization of Factors

We have just perceived how we can relate elective names, named aliases, with the tables that show up in the 'from' form. The name is utilized to allude to the table with regards to the query. This element can be used to maintain a strategic distance from the need for working out the full-name of the table every time it is asked for, as observed in Query-5 as well as for different reasons.

In any case, by utilizing assumed names we can allude more than once to the identical table, in a path like the utilization of the renaming administrator Q of relational algebra. Every time an alias is presented, we proclaim a variable of sort table, which has the substance of the table.

At the point when a table seems just once in an inquiry, there is no distinction between translating the 'alias' as a pseudonym or like another new variable. At the point when a table appears more than just once, it is fundamental to consider aliases to be new factors.

Query-17: Find every one of the workers who have similar surname (however unique to start with names) of a worker in the Production department.

select E_1.FirstName, E_1.Surname

from Employee E_1, Employee E_2

where E1.Surname = E2.Surname and

E_1.FirstName <> E_2.FirstName and

E_2.Dept = 'Production'

This question contrasts each line of EMPLOYEE and every line of EMPLOYEE related with the Production department. Note that in this query, each column whose Dept quality has the esteem Production is additionally contrasted and itself, yet the correlation of the line with itself will never be fulfilled, in that the predicate of imbalance on the trait FirstName can never be valid.

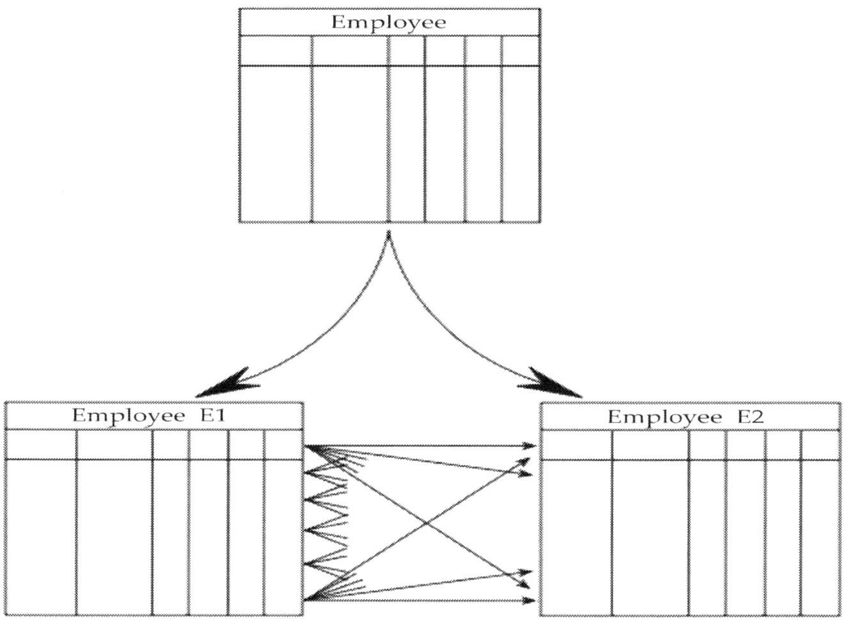

Figure 68: Description of Query 17.

(Source: Melton, J. S. Understanding the New SQL.)

To delineate the execution of this query, we can envision that when we characterize the false names, two distinct tables E1 and E2 are made, each containing every row of EMPLOYEE. Figure 68 demonstrates the thought and the way that there is an examination of each line of E1 with each line of E2.

The meaning of aliases is likewise critical for the determination of settled queries, as we will find in Section 4.2.6. To exhibit the correspondence between the renaming administrator of relational algebra and the utilization of factors in SQL, we can utilize SQL to plan the question appeared in Section 3.1.6 (Formula 3.3). Assume the table SUPERVISION(Head,Employee) depicts the connection between employees and heads.

Query-18: Find the name and pay rate of the head of the employees who earn more than 40 thousand.

```
select E1.Name as NameH,  E1.Salary as SalaryH
from Employee E1,  Supervision,  Employee E2
where E1.Number = Supervision.Head and
        E2.Number = Supervision.Employee and
        E2.Salary > 40
```

Requesting

Whereas a connection is comprised of a non-requested arrangement of tuples, applications regularly require a requesting of the lines. Consider the case in which the client needs to know the highest salary paid by the organization. This needs a query that profits the employee's information requested on the estimation of the Salary property.

SQL makes it conceivable to force a request on the columns of the consequence of a query through by clause. The syntax of is presented below.

order by *OrderingAttribute* [asc | desc]

{, *OrderingAttribute* [asc | desc]}

Along these lines, the query indicates the properties that must be utilized for the request. The lines are requested by the primary quality in the rundown. For lines that have a similar incentive for this characteristic, the estimations of the qualities that take after are considered in particular order. The request on each property can be ascending or descending determined by methods for the qualifiers asc or desc, respectively. In case the qualifier is not defined, the ascending request is taken. Consider the database in Figure 65.

Query-19: Extract the substance of the AUTOMOBILE table in ascending order of manufacturer (Make attribute).

```
select *
from Automobile
order by Make desc, Model desc
```

CarRegNo	Make	Model	DriverID
GHI 789	Lancia	Delta	PZ 1012436B
DEF 456	BMW	Z3	VR 2030020Y
ABC 123	BMW	323	VR 2030020Y
BBB 421	BMW	316	MI 202200330U

Figure 69: Query 19 Results.

(Source: Melton, J. S. Understanding the New SQL.)

Aggregate Queries

Total operations constitute a standout among the most essential expansions of SQL in examination with relational algebra.

In relational algebra, every one of the conditions is assessed one tuple at once. The condition is dependably a predicate that is assessed on each tuple independent of the others.

Frequently, it is important to assess properties that are reliant on sets of tuples. Assume that we wish to locate the quantity of employees in the Production department. The quantity of representatives will relate to the number of tuples in the connection EMPLOYEE having Production as the estimation of the trait Dept. This number, in any case, is not a property controlled by one tuple specifically, hence the inquiry is not expressible in relational algebra. Keeping in mind the end goal to express it in SQL, we utilize the total operation check.

Query-20: Find the number of employees in the Production department.

```
select count(*)
from Employee
where Dept = ‘Production’
```

The inquiry is propelled normally taking into account the ‘from’ and ‘where’ statements. The total oepration is then connected to the table containing the consequence of the inquiry. In the case above, a table is developed in the start containing every one of the real rows of EMPLOYEE having Production as the estimation of the quality Dept. At that point, the total operation is connected to the table, tallying the quantity of columns that show up in it. Standard SQL utilizes five total operations, namely count, sum, max, min, and avg.

The count operation has the following syntax.

count (<* | [distinct | all] *AttributeList*>)

The primary alternative (*) returns the number of rows. The distinct choice returns the number of various esteems for the characteristics in AttributeList. The all alternative returns the quantity of lines that have values unique in relation to the invalid incentive for the properties in AttributeList. In case that a value is not determined whether distinct or all, all is the default value.

Question 21: Find the number of various esteems on the quality Salary for all the actual rows in EMPLOYEE.

```
select count(distinct Salary)
from Employee
```

Question 22: Find the number of rows in EMPLOYEE having a not invalid incentive on the trait Salary.

```
select count(all Salary)

from Employee
```

Let us proceed with the next four 4 operations: sum, max, min, and avg. They permit as contention on quality or articulation, potentially went before by one of the watchwords, distinct or all. The sum and average (avg) permit as contention just articulations that speak to numeric esteems or interims of time. The maximum (max) and minimum (min) require just that a request be characterized in the articulation, making them material likewise to character strings and moments of time.

<sum | max | min | avg> ([distinct | all] *AttributeExpression*)

The operations apply to the rows that are chosen by the where provision of the query and have the following implications.

- sum returns the whole of the qualities controlled by the property articulation.
- max and min return individually the most extreme and least esteem.
- average restores the normal of the estimations of the quality articulation.

The keywords have the importance as we have examined: distinct disregards duplicates, all forgets just invalid esteems. The utilization of distinct or all with the max and min operations has no impact on the outcome. The different executions of SQL every now and again offer a more extensive scope of total operations, giving measurable operations like the difference, standard deviation, middle, and so on.

Query-23: Find the sum of the salary of employees in the Administration department.

```
select sum(Salary)
from Employee
where Dept = 'Administration'
```

We can likewise assess a few total operations in a solitary inquiry.

Query-24: Find the highest, the average, and the lowest salary of employees.

```
select max(Salary), avg(Salary), min(Salary)
from Employee
```

Total assessment may happen after a discretionary inquiry such as the below one, presented with a join.

Query-25: Find the highest salary among employees whose department is located in the city of London.

```
select max(salary)
from Employee, Department
where Dept = DeptName and
      Department.City = 'London'
```

Note that the following adaptation of a similar inquiry is not right.

Query-26:

```
select FirstName, Surname, max(Salary)
from Employee, Department
where Dept = DeptName and
      Department.City = 'London'
```

On a natural level, this query would choose the highest value of the Salary quality and along this would naturally choose the characteristics

FirstName and Surname of the comparing worker. In any case, such semantics could not be summed up to the total queries, for two reasons. In any case, there is no certification that the operation will choose a solitary component, given the fact that there might be more than one (1) row containing a specific pay. In the second place, if composed this way, the inquiry could be connected to the max and min operations but would have no significance for the other total operations.

Consequently, the SQL linguistic structure does not permit total capacities and characteristic articulations (such as, quality names) inside a similar target list (aside from query with by clause, portrayed in the following segment) (Melton J. , 1996).

Group by Queries

We have depicted the total operations as those that are connected to a set of rows. The cases that we have looked at so far work on every one of the rows that are created as the consequence of the inquiry. It is frequently important to apply a total operation to particular subsets of lines. To utilize the total operation like this, SQL has a group by clause that makes it possible to determine how to isolate the table into subsets. The statement acknowledges as an arrangement of traits and the query will work independently on each arrangement of lines that have similar esteems for this arrangement of traits. To better comprehend the importance of the group by clause, let us investigate an example shown below.

Query-27: Find the sum of the salary of employees in the same department. Assume that the table contains the data appeared in Figure 70.

```
select Dept, sum(salary)
from Employee,
group by Dept
```

EMPLOYEE

FirstName	Surname	Dept	Office	Salary	City
Mary	Brown	Administration	10	45	London
Charles	White	Production	20	36	Toulouse
Gus	Green	Administration	20	40	Oxford
Jackson	Neri	Distribution	16	45	Dover
Charles	Brown	Planning	14	80	London
Laurence	Chen	Planning	7	73	Worthing
Pauline	Bradshaw	Administration	75	40	Brighton
Alice	Jackson	Production	20	46	Toulouse

Figure 70: Table of EMPLOYEE.

(Source: Melton, J. S. Understanding the New SQL.)

To start with, the query is considered as though the group by clause did not exist, choosing the traits that seem either as the contention of group by or inside an articulation that is the contention of the total operation. In the inquiry that we are taking a gander at, it is as though the accompanying query will be executed as follows.

select Dept, salary

from Employee

The outcome is fully shown in Figure 71. The subsequent table is then examined, separating the rows into subsets having a similar incentive on the properties that show up in the group by clause.

In the illustration, the rows are assembled by a similar estimation of the property Dept (Figure 72). Once the rows are parcelled into subsets, the total operation is connected independently to every subset. The result of the query is shown in the table below.

Dept	Salary
Administration	45
Production	36
Administration	40
Distribution	45
Planning	80
Planning	73
Administration	40
Production	46

Dept	Salary
Administration	45
Administration	40
Administration	40
Distribution	45
Planning	80
Planning	73
Production	36
Production	46

Figure 71: Projection of Dept and Salary for the EMPLOYEE.

(Source: Melton, J. S. Understanding the New SQL.)

Dept	Salary
Administration	45
Production	36
Administration	40
Distribution	45
Planning	80
Planning	73
Administration	40
Production	46

Dept	Salary
Administration	45
Administration	40
Administration	40
Distribution	45
Planning	80
Planning	73
Production	36
Production	46

Figure 72: Regrouping of the Dept.

(Source: Melton, J. S. Understanding the New SQL.)

The table above has rows that contain the aftereffects of the assessment of the total operation, potentially close by the estimations of the characteristics utilized for the total. In Figure 73 the last consequence of the inquiry is appeared, giving the total of the salary paid to the employees per department.

Dept	Salary
Administration	125
Production	82
Distribution	45
Planning	153

Figure 73: Query 27 Results.

(Source: Melton, J. S. Understanding the New SQL.)

The SQL forces the limitation that at whatever point, the gathering by provision is utilized. The traits that can show up in the select statement must be a subset of the properties utilized as a part of the group by clause. References to these qualities are conceivable in light of the fact that each tuple of the group will be portrayed by the same values.

The case below illustrates the purposes behind this confinement.

Query-28:

```
select office
from Employee
group by Dept
```

This query is off base in that various estimations of the Office quality will relate to estimation of the Dept property. Rather, after the grouping has been completed, each subgroup of lines must compare to a single row in the table coming about because of the query.

Then again, this confinement can now appear to be unreasonable, like if when we have to demonstrate the estimations of properties that show single esteems for given estimations of the grouping traits. (We say that the properties practically rely upon the gathering traits).

Query-29:

```
select DeptName, count(*), D.City
from Employee E join Department D on (E.Dept = DeptName)
group by DeptName
```

This inquiry should restore the department name, the number of workers in every department, and the city in which the department is based. Given that the characteristic DeptName is the key of DEPARTMENT, a specific estimation of City compares to estimation of DeptName. The framework could hence give a right reaction, yet SQL restricts inquiries of this nature. In hypothesis, the language could be adjusted so as to permit the recognizable proof of the credits that are keys to the patterns of the tables and feature the traits that could show up in the objective rundown. Practically speaking, it is necessary to keep the language straightforward, conceivably requiring that the query utilize an excess grouping of characteristics. The inquiry can be effectively revamped as takes after:

Query-30:

```
select DeptName, count(*), D.City
from Employee E join Department D on (E.Dept = D.DeptName)
group by DeptName, D.City
```

Gathering Predicates

We have perceived how lines can be composed into subsets by methods for the group by clause. It is conceivable that application requirements to consider just the subsets that fulfil certain conditions. In the event that the conditions that must be fulfilled by the subsets are certain at the single push level, it is adequate to utilize the suitable predicates as the contention of the where clause. Assuming that the conditions need to allude to values that are the consequence of conglomerations, we need to utilize another build, the on clause.

The on clause portrays the conditions that must be connected toward the end of the execution of a query that uses the group by clause. Every subset of rows frames some portion of the aftereffect of the query just if the predicate contention of on clause is fulfilled.

Gathering predicates - We have perceived how lines can be sorted out into subsets by using the group by clause. It is conceivable that an application needs to consider just the subsets that fulfil certain conditions. In the event that the conditions that must be fulfilled by the subsets are certain at the single push level, it is adequate to utilize the suitable predicates as the contention of the where clause. Assuming that the conditions need to allude to values that are results of the sum, we need to utilize another that is the having clause.

The having clause depicts the conditions that must be connected toward the end of the execution of an inquiry that uses the group by clause. Every subset of lines shapes some portion of the consequence of the inquiry just if the predicate contention of having clause is fulfilled.

Query-31: Find which departments spend more than 100 thousand on salary.

```
select Dept
from Employee
group by Dept
having sum(Salary) > 100
```

Applying the query to the table appeared in Figure 70, we take after the means depicted for aggregate by queries. Once the columns have been assembled concurring to the estimation of the Dept trait, the predicate contention of the having clause is assessed. This chooses the departments for which the sum of Salary is greater than 100. The aftereffect of the inquiry is appeared in Query 4.24.

Dept
Administration
Planning

Figure 74: Query 31 Results.

(Source: Melton, J. S. Understanding the New SQL.)

The syntax additionally considers the meaning of inquiries utilizing the having clause, without a comparing bunch by condition. For this situation the whole arrangement of lines is dealt with as a solitary gathering, however this has extremely restricted application conceivable outcomes, since if the condition is not fulfilled, the subsequent table will be unfilled.

Like the where clause, the having clause will likewise acknowledge as contention a boolean articulation of basic predicates. The basic predicates are for the most part between the consequence of the assessment of a total operation and a nonexclusive articulation. The syntax additionally permits the nearness of predicates on the properties shaping the contention of the grouping by, yet it is desirable over gather all such basic conditions into the setting of the where clause. To set up which predicates of an inquiry that utilization gathering are contentions of the where clause and which are contentions of the having clause, we set up as it were the accompanying standard: just the predicates containing total operation ought to show up in the contention of the having clause.

Query-32: Find the departments in which the average salary of employees in Office 20 is greater than 25 thousand.

```
select Dept
from Employee
where Office = ‘20’
group by Dept
having avg(Salary) > 25
```

The general brief type of a SQL query along these lines progresses toward becoming:

SelectSQL ::=select *TargetList*

from *TableList*

[where *Condition*]

[group by *GroupingAttributeList*]

[having *AggregateCondition*]

[order by *OrderingAttributeList*]

Set Queries

Additionally, SQL gives set operations, like the one that we actually saw in relational algebra in the prior sections. The accessible operations are union, intersect, and except. Each has an importance like the social relational algebra operation.

Note that each query that uses union and except can be communicated utilizing different develops of the dialect (by and large utilizing settled questions, which we will examine in Section 4.2.6). Then again, for the same reasons talked about in Section 3.2.3 for relational analytics with run revelations, the union operation upgrades the expressive energy of SQL and permits the meaning of queries that would somehow be incomprehensible.

The syntax is depicted below.

SelectSQL {<union | intersect | except> [all] *SelectSQL*}

Set operations, dissimilar to whatever remains of the language, expect the disposal of copies as a default. We can see two purposes behind this. Right off the bat, the disposal of copies is substantially nearer to the instinctive semantics of these operations. Also, their execution (especially intersect and except) requires a calculation for which the extra cost of taking out copies would be insignificant. On the off chance that we wish to embrace an alternate understanding of the operations in the statement and we wish to utilize set operations that keep up the copies, it is adequate to indicate the all keywords when utilizing the operations. In the accompanying illustrations, we will think about what occurs in both cases. A further perception is that SQL does not require the outlines on which the actual operations are delivered to be indistinguishable. The correspondence between the traits is not in view of the name but on the position of the traits. On the off chance that the qualities have distinctive names, the outcome regularly utilizes the names of the main operand.

Query-33: Deliver the first name and surname:

```
select FirstName as Name
from Employee
    Union
select Surname
from Employee
```

The inquiry initially acquires the estimations of the characteristic FirstName in the EMPLOYEE. At that point, it acquires the estimations of the characteristic Surname for the same lines, and then builds the table by joining the two halfway outcomes. Given that the set operations take out the copies, there will be no rehashed components in the subsequent table regardless of the nearness of copies in both the underlying tables and some indistinguishable esteem in both tables. Taking the underlying information to be that of the table in Fig 70, the outcome of the assessment of the inquiry is appeared in Fig 75.

Name
Mary
Charles
Gus
Jackson
Laurence
Pauline
Alice
Brown
White
Green
Neri
Chen
Bradshaw

Figure 75: Query 33 Results .

(Source: Melton, J. S. Understanding the New SQL.)

Query-34: Find the first names and the surnames of the considerable number of employees, with the exception of those having a place with the Administration department, keeping the copies:

```
select FirstName as Name
from Employee
where Dept <> 'Administration'
        Union all
select Surname
from Employee
where Dept <> 'Administration'
```

For this situation every one of the copies are held. The consequence of the inquiry, once more in light of the table in Figure 70, is appeared in Figure 76.

Query-35: Find surnames that are the same as in the first names:

```
select FirstName as Name
from Employee
        intersect
select Surname
from Employee
```

From this inquiry, we acquire the outcome in Figure 77.

Query-36: Find first names that are not the same as in the surnames.

```
select FirstName as Name
from Employee
        except
select Surname
from Employee
```

Name
Charles
Jackson
Charles
Laurence
Alice
White
Neri
Brown
Chen
Jackson

Figure 76: Query 34 Results.

(Source: Melton, J. S. Understanding the New SQL.)

Name
Jackson

Figure 77: Query 35 Results.

(Source: Melton, J. S. Understanding the New SQL.)

Name
Mary
Charles
Gus
Laurence
Pauline
Alice

Figure 78: Query 36 Results.

(Source: Melton, J. S. Understanding the New SQL.)

Nested Queries

Up to now, we have taken a gander at queries in which the contention of the where clasue depends on compound limitations in which each of the predicates speaks to a straightforward correlation between two esteems. SQL also permits the utilization of predicates with a more unpredictable structure, in which an esteem articulation can be contrasted, and the aftereffect of the execution of a SQL query. The inquiry utilized for the correlation is characterized specifically in the interior predicate of the where clause and is known as a settled inquiry.

The esteem articulation to be utilized as the primary operand in the correlation is most normally a trait name. For this situation, we are then faced with the issue of heterogeneity in terms of the examination. On one side of the examination we have the aftereffect of the execution of a query in SQL, mostly an arrangement of qualities, while on the other side we have the estimation of the characteristic for the specific line. The arrangement offered by SQL comprises of utilizing the keywords all or any to expand the typical correlation operations (=, <>, <, >, <=, and >=). The keyword determines that the row fulfills the condition in case that the examination predicate of the quality incentive for the row is valid with a minimum of one of the components returned by the inquiry. The language structure requires that the space of the components returned by the settled inquiry is perfect with the characteristic with which the correlation is made.

Query-37: Find the employees who work in the different departments in London:

```
select FirstName, Surname
from Employee
where Dept = any (select DeptName
from Department
where City = 'London')
```

The inquiry chooses the lines of EMPLOYEE for which the estimation of the quality Dept is equivalent to no less than one of the estimations of the DeptName characteristic of the lines of DEPARTMENT with City equivalent to London.

This inquiry can likewise be communicated utilizing a join between the tables EMPLOYEE and DEPARTMENT, and the streamlining agents are by and large ready to manage the two distinct definitions of this inquiry similarly. The decision between one portrayal and the other can be impacted by the level of clarity of the arrangement. In cases as basic as this, there is no distinction, for more mind boggling inquiries, the utilization of a few settled queries can make strides decipherability.

Let us consider a query that enables us to discover the employees in the Planning department who have the same first name with the employees in the Production department. The question fits both conceivable plans: the first is smaller and utilizes factors.

Query-38:

```
select E1.FirstName, E1.Surname
from Employee E1, Employee E2
where E1.FirstName = E2.FirstName and
    E2.Dept = ‘Production’ and
    E1.Dept = ‘Planning’
```

The second strategy utilizes a settled question, communicating the inquiry without requiring aliases:

Query-39:

```
select FirstName, Surname
from Employee
where Dept = ‘Planning’ and
    FirstName=any (select FirstName
    from Employee
    where Dept = ‘Production’)
```

We think now about an alternate question.

Query-40: Find the departments in which there is nobody named Brown.

```
select DeptName
from Department
where DeptName <> all (select Dept
```

```
from Employee
where Surname = 'Brown')
```

The settled inquiry chooses the estimations of Dept for every one of the lines in which the surname is Brown. The condition is in this way fulfilled by the lines of department for which the estimation of the characteristic DeptName is not among the values created by the settled inquiry. This inquiry could nt be communicated by methods for a join. Note how this inquiry could have been actualized in relational algebra utilizing the articulation below (πDeptName(DEPARTMENT)−QDep tName←DeptπDept(σSurname=›Brown›(EMPLOYEE))), and in this manner could likewise have been communicated utilizing the except clause.

Query-41:

```
select DeptName
from Department
        except
select Dept as DeptName
from Employee
where Surname = 'Brown'
```

SQL offers two exceptional operations to represent set participation and its invalidation, in and not in. These are to be sure shorthand for = any and <>, respectively. Cases of their utilization are given in the questions of the following area. We can watch how a few queries utilize the max and min operations and could be communicated by methods for settled inquiries.

Query-42: Find the department of the employees having the highest salary (utilizing the max operation):

```
select Dept
from Department
where Salary = any (select max(Salary)
from Employee)
```

Query-43: Find the department of the employees having the highest salary (utilizing just a settled query):

```
select Dept
from Department
where Salary >= all (select Salary
from Employee)
```

The two queries are comparable in that the most extreme esteem is precisely the esteem that is the greater than or equal to every one of the estimations of a similar characteristic in alternate lines of the connection. In this case, it is fitting to utilize the total operation, as it gives a more comprehensible outcome (and conceivably more proficient). Additionally, it is fascinating to take note that for the first settled query there is no contrast between utilizing the keyword all, since the query dependably restores a solitary line.

Complex Settled Queries

A basic and instinctive elucidation for understanding settled queries lies in the presumption that the settled inquiry is executed before the investigation of the lines of the outside question. The consequence of the query can be spared in an impermanent variable and the predicate of the outside query can be assessed by utilizing the transitory outcome. Moreover, this understanding produces a proficient execution, since the settled query is prepared just once. Consider again Query-40. The system would first be able to do the settled question, which restores the estimation of the Dept trait for every one of the employees named Brown. When this is done, every division is verified that the name is excluded in the table created, utilizing the operator <> all.

Then, the settled query alludes to the setting of the inquiry in which it is encased. This occurs by methods for a variable characterized inside the outer inquiry and utilized as a part of the interior question. Such a system is regularly portrayed as the 'exchange of ties' starting with one setting then onto the next. This system improves the expressive energy of SQL. For this situation, the basic translation given before for settled queries is not any more substantial. Rather, it is important to re-evaluate the standard understanding of SQL inquiries, which initially figures the Cartesian result of the tables in the 'from' condition and next assess the where provision on each row. The settled question is a segment of the where clause and it should likewise be assessed independently for each line delivered by the thought of the outside inquiry.

Therefore, the new elucidation is as follows. For each row of the outside inquiry, we firstly assess the settled question then process the line level predicate of the outer query. Such a procedure can be self-assertively settled for a self-assertively complex utilization of factors inside the settled queries. In any case, the decipherability and explanatory nature of the language is traded off. With respect to the perceivability (or degree) of SQL factors, there is a confinement on the utilization of a variable. A variable can be utilized just inside the query in which it is characterized or inside an inquiry that is recursively settled in the query where it is characterized.

In the event that an inquiry contains settled questions at a similar level (on particular predicates), the factors presented in the 'from' clause of a question cannot be utilized as a part of the setting of another query. The accompanying question, for instance, is mistaken.

Query-44:

```
    select *
from Employee
where Dept in (select max(Salary)
    from Department D1
    where DeptName = 'Production') or
    Dept in (select DeptName
    From Department D2
    Where D1.City = D2.City)
```

We represent the importance of complex settled inquiries together with the depiction of another consistent operation, exists. This operator permits a settled question as a parameter and returns the genuine esteem just if the inquiry does not deliver a vacant outcome. This opeator can be conveniently utilized just if there is a coupling exchange between the outside inquiry and the settled inquiry. Consider a table PERSON (TaxCode, FirstNames, Surnames, City) portraying individuals' assessment records.

Query-45: Find every one of the homonyms, that is, people who have a similar first names and surnames yet unique assessment codes:

```
select *
from person P
where exists (select *
              from Person P1
              where P1.FirstName = P.FirstName and
              P1.Surname = P.Surname and
              P1.TaxCode <> P.TaxCode)
```

The inquiry looks through the lines of the PERSON-table for which there actually exists an additionally push in PERSON with the same FirstNames and Surnames, yet an alternate TaxCode.

For this situation, we cannot execute the settled inquiry before assessing the outside inquiry, given that the settled query is not appropriately characterized until a esteem has been doled out to the variable P. It is essential, rather, to assess the settled question for every line created inside the outside inquiry. Hence, in the illustration, the lines of the variable P will initially be inspected one by one. For each one of these lines, the settled query will be delivered and executed and will return or not the vacant outcome relying on regardless whether there are people with a similar names and surnames. This inquiry could likewise have been detailed with a join of the PERSON table with itself.

Look now at an alternate query.

Query-46: Find all people who do not have homonyms:

```
select *
from person P
where not exists (select *
from Person P1
where P1.FirstName = P.FirstName and
P1.Surname = P.Surname and
P1.TaxCode <> P.TaxCode)
```

The understanding is like that of the first question, with the single distinction that the predicate is fulfilled if the settled inquiry restores a discharge result. This query could likewise have been actualized by a distinction that subtracted from all the first names and surnames, those of the general population sharing a first and second name, decided utilizing a join. Another approach to define a similar query utilizes the tuple constructor, spoken to by a couple of bended sections that encase the rundown of traits.

Query-47: Find every one of the people who do not have homonyms (utilizing a tuple constructor)

```
select *
from person P
where (FirstName, Surname) not in (select FirstName, Surname
from Person P1
where P1. TaxCode <> P. TaxCode)
```

Consider the database with an actual table SINGER(Name,Song) and a table SONGWRITER(Name,Song).

Query-48: Find the artists who have performed just their own particular melodies:

```
select Name
from Singer
where Name not in (select Name
from Singer S
where Name not in
(select Name
From Songwriter
Where Songwriter.Song = S.Song))
```

The outside question has no connection with the settled inquiries, and in this manner can be at first suspended, sitting tight for the aftereffect of the principal level settled inquiry.

Such first-level query, presents an official-binding. Along these lines, the inquiry is executed by methods for the accompanying stages.

- The settled question select Name from Singer S is connected to every row of the table Singer.

- For each one of the rows, the most internal inquiry is assessed. This profits the names of the musicians of the melody titles showing up in the lines of S. Hence, if the name of the vocalist does not show up among the names of the musicians (which in this manner implies that the vocalist is not the essayist of the tune he sings), the name is chosen.
- Lastly, the table that contains the names of the vocalists who have performed just their own particular tunes (the more outside inquiry) is developed. This inquiry restores every one of the lines who's Name does not show up as aftereffect of the settled queries.

The rationale of this inquiry execution is clearer when we consider that the query can be communicated similarly by utilizing the except operator:

Query-49:

```
select Name
from Singer
except
select Name
from Singer S
where Name not in (select Name
from Songwriter
where Songwriter.Song = S.Song)
```

Business SQL systems do not generally do the settled questions inside by filtering the outside table and delivering an inquiry for each line of the connection. Rather, they attempt to process whatever number of questions as would be prudent in a set-arranged manner, with the point of dealing with an expansive amount of information by methods for as couple of operations as could reasonably be expected. A few enhancements are conceivable, for example, holding the aftereffects of the settled sub-queries for numerous utilizations and foreseeing the assessment of the parts of inquiry predicates that are definitely not settled.

DATA MODIFICATION IN SQL

The Data Manipulation Language (DML) of SQL incorporates orders for querying and modifying the database. The charges that permit the adjustment of the substance of a database are insert, delete, and update. We will break down the singular orders independently, in spite of the fact that, as we should see, they have comparative shapes.

Insertions

Summon for inserting a row into the database has the syntax that follows.

insert into *TableName* [(*AttributeList*)] <values (*ListOfValues*) | *SelectSQL*>

The main shape permits the addition of single row into the tables. The contention of the qualities statement speaks to unequivocally the estimations of the traits of the single row. For instance:

```
insert into Department (DeptName, City)
                values('Production', 'Toulouse')
```

The second frame permits the expansion of an arrangement of row, which are first recovered from the database.

```
insert into LondonProducts
        (select Code, Description
from Product
where ProdArea = 'London')
```

Above, results to the considerable number of rows having London as the estimation of the quality ProdArea, into the table LondonProducts. The two structures have diverse applications. The first case is regularly utilized in projects to fill in a table with information given by the clients. Each utilization of the insert statement is connected with the filling of a frame, that is, an inviting interface in which the client supplies esteems for different traits. The second case inserts information into a table in light of other data effectively shown in the database.

In the event that the estimations of a portion of the properties of the table are not indicated amid an addition, the default esteem is allocated to them, or coming up short this, the invalid esteem (Part 4.1.6). On the off chance that the addition of the invalid esteem disregards a not invalid imperative characterized on the characteristic, the addition will be rejected. At long last, take note of that the correspondence between the characteristics of the table and the qualities to be inserted is directed by the request in which the terms show up in the meaning of the table. Hence, the primary component of the ValuesList (for the principal type of the announcement) or the main component of the objective rundown (for the second shape) must relate to the main property that shows up in AttributeList (or in the meaning of the table if AttributeList is overlooked), and so on for alternate traits.

Deletions

The delete statement takes out rows from the actual tables of the database, which has the following syntax.

delete from *TableName* [where *Codition*]

At the point when the condition framing the subject of the where clause is not defined, the statement expels every one of the lines from the table, generally as it were the lines that fulfil the condition are erased. On the off chance that there is a referential trustworthiness limitation with a course approach in which the table is referenced, the cancelation of rows of this table can actually cause the cancelation of lines having a place with different tables. This can produce a chain response if these cancelations in their turn cause cancelation of rows in different tables.

delete from Department

where DeptName = 'Production'

Above statement erases the line of the Department having the names of the Production. (Since DeptName actually was proclaimed as the essential key for the table, there can be just a single row having that specific esteem.) The condition has an indistinguishable syntax as in the select clause, which implies that settled questions alluding to different tables can show up inside it. A basic case is the charge that deletes records where there is no employee in the department.

delete from Department

where DeptName not in (select Dept form Employee)

Note the contrast between the delete statement and the drop order portrayed in Section 4.1.8. A charge, for example,

delete from Department

erases every one of the lines of the DEPARTMENT table, at the same time erasing all the lines of the tables that are connected by referential honesty requirements with the table if the course strategy is indicated in case of cancellation.

Updates

The update statement is marginally more complex. The charge:

update TableName

set *Attribute* = <Expression | *SelectSQL* | null | default>

{, *Attribute* = <Expression | *SelectSQL* | null | default >}

[where *Condition*]

The update statement makes it conceivable to refresh at least one trait of the lines of TableName that fulfil a conceivable condition. On the off chance that the condition does not show up, a genuine esteem is accepted as default, and the change is conveyed out on every one of the rows. The new incentive to which the quality is appointed can be one of the following:

- the aftereffect of the assessment of an articulation on the traits of the table;
- the aftereffect of a SQL inquiry;
- the invalid esteem; or
- the default an incentive for the space.

The statement:

update Employee set Salary = Salary + 5

where RegNo = 'M2047'

works on a solitary line, refreshing the compensation of the worker number m-2047 though the accompanying illustration works on an arrangement of rows.

update Employee set Salary = Salary * 1.1

where Dept = 'Administration'

The charge delivers a 10% compensation increment for every one of the employyes who work in Administration. The task administrator has the typical property, for which the Salary on the right-hand side of the operator speaks to the old esteem, which is assessed for each row to which the refresh must be connected. The consequence of the articulation turns into the new estimation of the pay. The set-situated nature of SQL ought to be considered when composing update statement. Assuming that we wish to alter the employee's salary, increasing the salary under 30 thousand by 10 %, and the higher salary by 15 %. One may approach to initiate two update statements to achieve the objective as presented below.

```
update Employee set Salary = Salary * 1.1
where Salary < = 30
update Employee set Salary = Salary * 1.15
where Salary > 30
```

There will be a problem with the solution presented above. The first statement will update those records in which the salary of employee amounts to 30 thousand and less. Hence, those records in which the employee's salary is 30 thousand will become 30.8 thousand. In this case, the second update statement will modify the remaining records not modified in the first update statement including those records modified after initiating the first update statement in which the updated salary reached to more than 30 thousand like in aforementioned case.

The issue starts on the grounds that SQL is set-situated. With a tuple-arranged language it is conceivable to refresh the rows one by one and apply one or, on the other hand the other adjustment as per the estimation of the pay. In this specific case, the arrangement is to upset the request of execution of the two statements, first expanding the higher pay rates and after that the staying ones. In more perplexing circumstances, the arrangement may require the presentation of middle of the road refreshes, or a total change of approach utilizing a program in an abnormal state programming language, utilizing cursors.

OTHER MEANINGS OF INFORMATION IN SQL

Having portrayed how to plan queries in SQL, we can finish the rundown of the segments of an outline. We will now portray the check clause, assertion, and view.

Generic Respectability Imperatives

We have perceived how SQL permits the determination of a specific arrangement of imperatives on the properties and tables, fulfilling the most essential, yet not all of the application prerequisites. For the particular of further requirements, SQL - 2 has presented the check statement, with the below syntax:

The conditions that can be utilized are those that can show up in the where clause of an SQL statement. The condition should dependably be confirmed to keep up the rightness of the database. Along this line, it is conceivable to determine all the tuple limitations we talked about in Section 2.2.1, and much more, on the grounds that the Condition can incorporate references to different fields. A viable exhibition of the energy of the build is to indicate how the predefined requirements would be portrayed by using the check clause.

We can rethink the pattern of the primary adaptation of the EMPLOYEE table given in Section-4.1.7.

Check *Condition*

Create table Employee

```
(
RegNo character(6)
check (RegNo is not null and
1 = select count(*)
from Employee E
where RegNo = E.RegNo))
Surname character(20) check(Surname is not null)
Name character(20) check(Name is not null and
1 = select count(*)
from Employee E
where Name = E.Name
and Surname = E.Surname))
Dept character(15) check (Dept in
```

(select DeptName
from Department))
)

By contrasting this particular and the one utilizing predefined limitations, we can mention various objective facts. In the first place, the predefined imperatives permit a more minimized and comprehensible portrayal. For instance, the key imperative requires a genuinely complex portrayal, which utilizes the total operation count. Note likewise that utilizing the check clause, we lose the likelihood of relegating a strategy of response to infringement to the limitations. Lastly, when the imperatives are characterized utilizing the predefined develops, the framework can remember them quickly and can check them all the more productively.

To comprehend the energy of the new check develop, we can, for instance, depict an imperative that strengths the worker to have his very own supervisor division just if the enrolment number does not start with the esteem 1.

The above table can be reached out with the following statement.

Superior character(6)
check (RegNo like "1%" or)
Dept = (select Dept
from Employee E
where E.RegNo = Superior))

Assertions

Other than the check condition, we can characterize a further segment of the diagram of the database through assertion. Presented in SQL - 2, this speaks to requirements that are not related with any row or table specifically, and have a place with the pattern.

Utilizing this statement, we can express every one of the imperatives that we have as of now determined when managing table definitions. In any case, assertion also permits the articulation of limitations that would not generally be expressible, such as limitations on more than one table or imperatives that require a table having a base cardinality. Declarations are named, and accordingly they can be expressly dropped though drop in the alter statement (see Segment 4.1.8).

The syntax for the use of assertion is presented below.

Create assertion *AssertionName* check (*Condition*)

For instance, a declaration can force that in EMPLOYEE there is dependably at slightest one row exhibit.

```
create assertion AlwaysOneEmployee
check (1 <= (select count(*)
from Employee))
```

Each trustworthiness limitation, either check or assertion, is related with a check strategy that shows whether it is prompt or conceded. Quick imperatives are checked instantly after every change of the database, while conceded imperatives are checked just toward the finish of a progression of operations, called an exchange. At the point when a quick imperative is not fulfilled, it is on the grounds that the particular information change operation causing the infringement was quite recently executed and it can be 'fixed'; this is called incomplete rollback. All the settled arrangement imperatives, presented in 4.1.6 (not invalid, one of a kind, essential key) and 4.1.7 (remote key), are instantly checked, and their infringement causes a halfway rollback. At the point when an infringement of a conceded imperative is distinguished toward the finish of an exchange, the information adjustment operation causing the infringement is not known, and hence the whole exchange must be fixed; this is called rollback.

Because of the above systems, the execution of an information control guideline on a database occasion that fulfils all the requirements will dependably create a database occasion that likewise fulfils all the imperatives. We likewise say that they deliver a 'predictable database state'.

Inside a program, we can change the quality of an imperative, by setting it quick or conceded; this is finished by methods for the summons set imperatives [ConstraintName] quick and set limitations [ConstraintName] conceded.

Views

During Section 3, views were presented as "virtual" tables, whose substance relies upon the substance of alternate tables in the database. Views are characterized in SQL by partner a name and a rundown of qualities with the aftereffect of the execution of a query. Different views can likewise show up inside the inquiry that characterizes the view. Nonetheless, views cannot be commonly reliant, either instantly (characterizing a view as far as itself), or transitively (characterizing a view V1 utilizing a view V2, V2 utilizing V3 ... until the point when Vn is characterized by implies of V1),

A view is characterized utilizing the following statement.

create view *ViewName* [(*AttributeList*)] as *SelectSQL*

[with [local | cascade] check option]

The SQL query and the mapping of the view must have a similar number of qualities. We can, for instance, characterize a view ADMINEMPLOYEE that contains every one of the employees in the Administration department with salary higher than 10-thousand.

```
    create view AdminEmployee(RegNo, FirstName, Surname, Salary)
as
    select RegNo, FirstName, Surname, Salary
    from Employee
where Dept = 'Administration' and
Salary > 10
```

We would now be able to develop a view JuniorAdminEmployee in light of the AdminEmployee actual view, which would contain the regulatory employees with a salary between 10-thousand and 50-thousand:

```
    create view JuniorAdminEmployee as
select *
from AdminEmployee
where Salary < 50
with check option
```

On specific view, we can complete change operations. This must be converted into the fitting alteration orders on the construct tables with respect to certain view. As we have just specified in Section 3, we cannot generally locate an unambiguous method for altering the base table or tables. Troubles are experienced, especially when the view is characterized by methods for a join between more than one table. In the standard SQL, a view can be refreshed just when a solitary line of each base table compares to a row in the view.

Business SQL systems think about that as a view can be refreshed just in the event that it is characterized on a solitary table. Few frameworks require that the characteristics of the view ought to contain no less than an essential key of the table. The check alternative condition can be utilized just with regards of views that can be refreshed.

It determines that update operations can be completed just on lines having a place to the view, and after updating the lines must keep on belonging to this view. This can happen, for instance, if esteem is relegated to a quality of the view that actually makes one of the determination predicates false. At the point when a view is characterized as far as different perspectives, the nearby or fell alternative indicates regardless of whether the control over row expulsion must be made just at the (nearby) see or it must be spread to every one of the perspectives on which the view depends. The default alternative felled.

Since the JuniorAdminEmployee view has been characterized utilizing the check, an endeavour to relegate an estimation of 8-thousand to the Salary characteristic would not be acknowledged by the present meaning of the view, but would be acknowledged if the check choice were characterized as neighbourhood. An alteration of the SaLaryquality of a line of the view to dispense the esteem 60-thousand would not be acknowledged even with the nearby choice (Codd E. F., 1970).

Views in Queries

Views can be utilized as a part of SQL to plan queries that in some way or another is difficult to express. Utilizing the meaning of proper views, we can characterize inquiries in SQL that require a settling of different total operations or make a complex utilization of the union operation. Views can be considered as an apparatus that builds the likelihood of making settled questions.

We wish to figure out which office has the highest salary. Keeping in mind the end goal to accomplish this, we characterize a view that will be utilized by Inquiry 50.

create view SalaryBudget (Dept, SalaryTotal) as

select Dept, sum(Salary)

from Employee

group by Dept

Query-50: Find the department with the highest salary.

select Dept

from SalaryBudget

where SalaryTotal = (select max(SalaryTotal)

from SalaryBudget)

The meaning of the view SalaryBudget constructs a table in which there shows up a line for each department. The ascribe Dept compares to the trait Dept of EMPLOYEE and contains the name of the department. The second trait SalaryTotal contains, for every department, the total salary of certain number of employees of the said department.

Another approach to figure a similar query is as follows.

Query-51:

select Dept

from Employee

group by Dept

having sum(Salary) >= all (select sum(Salary)

from Employee

group by Dept)

This arrangement is not perceived by all SQL systems frameworks, which may force the condition in the having clause must be a straightforward examination with a quality or a steady and cannot utilize a settled inquiry. We can give another case of the utilization of views to detail complex queries.

create view DeptOffice(DeptName,NoOffices) as

select Dept, count(distinct Office)

from Employee

group by Dept

Query-52: Find the normal number of offices per department.

Select avg(NoOffices)

From DeptOffice

Query-53:

select **avg(count**(distinct Office))

from Employee

group by Dept

The query is nonetheless mistaken on the grounds that the SQL language structure does not permit the falling of total operations. The essential issue is that the assessment of the two distinct operations occurs at various levels of total, while a solitary event of the group by clause for each inquiry is acceptable.

ACCESS CONTROL

The nearness of information insurance systems goes up against awesome significance in present day applications. A standout among the most vital undertakings of a database overseer is the decision and execution of fitting access control strategies. SQL perceives the significance of this perspective and an arrangement of guidelines is committed to this capacity.

SQL is composed with the goal that each client is identifiable by the framework as it is unambiguous. The client recognizable proof can abuse the abilities of the working framework (so every client of the framework relates to a client of the database) or database clients can be free of system clients. Most business systems offer a free association, each with its own distinguishing proof technique. Along these lines, more than one framework client can relate to a database client and the other way around.

Resources and Benefits

The assets that are ensured by the systems are typically tables. However, we can ensure any of the framework segments, such as the properties of a table, views, and domains.

When in doubt, the client who makes an asset is its proprietor and is approved to make any operation upon it. The system in which just the proprietors of the assets were approved to make utilization of them would be of constrained use, as would a system in which every one of the clients could utilize each asset in a subjective way. SQL offers adaptable association instruments that enable the director to indicate those assets to which the clients may approach and those that should be kept privately. The system bases get to control on an idea of benefit. The clients have benefits of access to the system assets.

Each benefit is described by the following:

- the asset to which it alludes;
- the client who allows the benefit;
- the client who gets it;
- the activity that is permitted on the asset; and
- regardless of whether the benefit can be passed on to different clients.

At the point when an asset is made, the system consequently surrenders all the benefits on that asset to its maker. Furthermore, there is a predefined client, which speaks to the database head, who has all benefits on all assets.

The accessible benefits are as per the following:

- embed: permits the inclusion of another protest into the asset (this can be connected just to tables or perspectives);
- refresh: enables the estimation of a protest be refreshed (this can be utilized on tables, views, and traits);
- erase: permits the expulsion of a question from the asset (tables or perspectives as they were);
- select: enables the client to peruse the asset, keeping in mind the end goal to utilize it inside a query (utilized on tables, views, and qualities);
- references: enables a reference to be made to an asset with regards to the meaning of a table. It can be connected just with tables and particular traits. With the references benefit (say on the table DEPARTMENTs, property of Paolo), a client who is allowed the benefit (say, Stefano) is ready to characterize a

remote key limitation (for instance, on his table EMPLOYEE) referencing the asset that is allowed (for instance, the key of Office). Now, if Stefano determines a no activity approach on the referential uprightness requirement, Paolo is kept from crossing out or altering lines of his own table DEPARTMENT, if such a refresh order renders the substance of EMPLOYEE mistaken. In this way, giving out a references benefit on an asset may restrict the capacity to alter that asset; and

- utilization: applies to areas, and enables them to be utilized, for instance, in the meaning of the pattern of a table.

The benefit of completing a drop or modify cannot be fully completed without a doubt, but remains the privilege of the maker of the protest itself. Benefits are yielded or transmitted by methods for the directions allow and renounce.

Commands for Grant and Revoke Privileges

The syntax of the grant command is the following.

grant *Privilege* on *Resource* to *Users* [with grant option]

The order permits the allowing of Privileges on the Resource to the Clients. For instance, the order:

grant select on Department to Stefano

grants client Stefano the select benefit on the DEPARTMENT table. The 'with' allow choice provision indicates whether the benefit of proliferating the benefit to different clients should likewise be allowed. We can utilize the keywords all privileges set up of the unmistakable privileges. These distinguish every one of the privileges that the client can allow on a specific asset. Along these lines the order:

grant all Privileges on Employee to Paolo, Riccardo

surrenders to the clients Paolo and Riccardo every one of the privileges on the EMPLOYEE table that can be granted by whoever executes the command.

The revoke command does the opposite, which is taking away the privileges that a client had just been allowed.

revoke *Privileges* on *Resource* from *Users* [restrict | cascade]

Among the privileges that can be revoked, aside from those that can show up as subject of the grant command, is the grant choice benefit that gets from the utilization of the condition 'with' allow alternative.

The client who can revoke privileges from another client is the client who granted the privileges on the first place. A revoke can take away all or a subset of the privileges that were initially granted. The default alternative limit indicates that the order must not be completed on the off chance that revoking this client's privileges causes a further denying of benefits. This can happen at the point when the client has gotten the benefits with the give choice and has engendered the benefits to different clients. In a circumstance of this nature, the confine alternative will make a blunder be flagged. With the course alternative, rather, the execution of summon is basic. Every one of the benefits that were proliferated are repudiated and every one of the components of the database that were developed by abusing these benefits are evacuated. Note that for this situation as well, the course choice can produce a chain response, in which for each component expelled, every one of the items that have some needy connection to it are additionally evacuated. As in other similar cases, it is important to be exceptionally watchful that a straightforward command does not create broad and undesirable alterations on the database.

It is not just the revoke command can deliver chain responses. The cascade-concede order can likewise create comparative impacts. It can happen that a client has gotten a benefit on a table that has enabled him to make views that allude to this table, by methods of, for example, the select benefit. On the off chance that the client is conceded additionally benefits on the table, at that point these benefits will be naturally stretched out to cover the perspectives (and along these lines to views built on these perspectives).

SQL IN PROGRAMMING LANGUAGES

It is uncommon that entrance to data contained in a database occurs as a consequence of individual collaboration utilizing SQL. The most run of the mill utilization of a database occurs by methods for coordinated applications incorporated with the data framework, while coordinate utilization of the SQL translator is held for a few master clients.

The utilization of unique applications instead of the SQL translator to access data is legitimized by various components. All the time, access to the data is not required straightforwardly by a client but by a non-intelligent (cluster) application. Further, for intuitive clients, the get strategies are regularly basic. It is in this manner helpful to lessen the many sided quality of access to the database by developing an application that gives an improved interface for doing the errand. Lastly, the introduction of the information offered by the framework may be unsatisfactory for the client's prerequisites, while an uncommon application is free from imperatives and can give a portrayal satisfactory for the necessities (Melton J. , 1996).

There are many devices that can be utilized in making database applications. A flourishing business sector is that of the fourth generation languages (4gls), a very refined advancement instruments that make it conceivable to create finish database administration applications with little exertion. Most database sellers offer, together with the database director, an arrangement of apparatuses for the advancement of utilizations. In addition, there is a rich supply of items that are not connected to any database, specifically all of which can deal with an exchange with the relational algebra through standard SQL.

These apparatuses make conceivable the successful meaning of database mappings and the development of complex interfaces.

Another strategy for composing applications utilizes conventional abnormal state programming languages. We will focus our examination on this strategy, since it is still of significance and as a result of the absence of consistency among the 4gls.

Integration Issues

So as to utilize SQL directions inside a procedural program, the SQL guidelines must be embodied inside the program. From the execution perspective, it is important to give the abnormal state language compiler with a pre-processor. This pre-processor can recognize the calls to the DBMS benefits and submit them to the DBMS inquiry execution condition, which incorporates the inquiry analyser. This arrangement offers the common focal points of convenience and deliberation that as of now describe the utilization of a standard language such as SQL. At execution, the program starts a discourse with the database, sending the inquiry specifically to the framework.

One specific issue lies in the way that programming languages get to the components of a table by checking the lines one by one, utilizing what is known as a tuple-arranged approach. Conversely, SQL is a set-situated language that dependably follows up on whole tables and not upon single lines.

Indeed, the result of an SQL query is a whole table. This view causes the issue known as impedance confound.

There are two conceivable answers for this issue. The first comprises of utilizing a programming language that makes more capable information builds accessible, specifically can sort out an 'arrangement of lines' sort structure in a whiz way. This arrangement is increasing more enthusiasm because of the expanding spread of queries situated programming languages, portrayed by intense type-definition systems. Notwithstanding, most applications are composed utilizing languages that do not have this capacity. A further trouble with this arrangement is the absence of a standard arrangement acknowledged by all frameworks, such as by protest arranged and question social frameworks.

The second system that is more normally utilized, has been institutionalized and does not require a muddled expansion of existing programming dialects, despite the fact that it is not as abnormal state and benevolent to the software engineer as one would like it to be. This arrangement depends on the utilization of cursors.

Cursors

Cursors are components that enable a program to get to the rows of a table each one in turn. It is characterized by utilizing an inquiry.

Declare *CursorName* [scroll] cursor for *SelectSQL*

[for <read only | update [of *Attribute* {, *Attribute*}]>]

The pronounce cursor charge characterizes a cursor related with a specific query on the database. The parchment alternative determines whether we wish to permit the program to move unreservedly on the aftereffect of the inquiry. The last alternative for refresh indicates whether the cursor can be utilized for communicating an update command, conceivably including the detail of the traits that is destined to be refreshed.

Open *CursorName*

The open command applies to a cursor. When conjured, the inquiry is conveyed out and the inquiry result can be gotten to utilizing the bring order.

fetch [*Position* from] *CursorName* into *FetchList*

The get command takes a full row from the actual cursor and returns its esteems into the factors of the program that show up in FetchList. The FetchList must incorporate a variable for each component in the objective rundown of the query, with the end goal that each component of the FetchList is sort perfect with the areas of the components of the objective rundown of the question. An imperative idea is that of current rows, which speaks to the last line utilized by the bring operation. The Position parameter is utilized to set the present row. It can accept the following values:

- next (the present line turns into the line that takes after the present one);
- first (the principal line of the inquiry result);
- last (the last column);
- outright IntegerExpression (the line that shows up in the i-th position of the question result, where i is the aftereffect of the assessment of the articulation);
- relative IntegerExpression (the line that shows up in the i-th position, beginning from the present position).

These alternatives can be utilized just on condition that the parchment choice (which ensures the likelihood of free development inside the aftereffect of the inquiry) was determined when the cursor was characterized. In the event that the parchment choice is most certainly not determined, the main esteem accessible for the parameter is next. For this situation, execution is simpler as the lines of the outcome can be disposed of promptly in the wake of being returned. This can be extremely valuable for diminishing reaction times, especially when the question restores an extensive number of rows.

update *TableName*

set *Attribute* = <*Expression* | null | default>

{, Attribute = <Expression | null | default>}

where current of *CursorName*

The update and delete commands enable alterations to be made from database by methods of cursors. The main augmentation required to the language structure of update and delete lies in the likelihood of utilizing as a part of the where clause, the keyword CursorName shows the present rows (that must be refreshed or expelled). The alteration charges can be utilized as it were in the event that the cursor offers access to a real line of a table and is not relevant when the question related with the cursor requires a join among various tables.

close *CursorName*

Now, the assets committed to the cursor are discharged, especially the memory space putting away the outcome.

A basic case is as shown below.

```
declare EmployeeCursor scroll cursor for
select Surname, Firstname, Salary
from Employee
where Salary < 100 and Salary > 40
```

Along this line, the cursor EmpoloyeeCursor is dispensed to the query that makes it conceivable to acquire information identifying with the representatives who procure between 40-thousand and 100-thousand.

A basic case of a C-program that utilizes cursor is explained below. The SQL charges are recognized by the character "$" in the main section, and the factors of the program are recognized by the way that their names are gone before by the colon (:) character. The factors must be of a sort with the qualities that they will contain. A predefined variable, SQL-code, is utilized. It actually contains 0 if the development of the last order has been fruitful and a non-zero blunder code generally. Its primary utilization is to distinguish when rows of the cursor have all been brought.

```
void DisplayDepartmentSalaries(char DeptName[])
{
        Char FirstName[20], Surname[20];
        Long int Salary;
$ declare DeptEmp cursor for
```

```
    select FirstName, Surname, Salary
from Employee
where Dept = :Deptname;
$ open DeptEmp;
$ fetch DeptEmp into : FirstName, :Surname, :Salary;
Printf("Department %s\n",DeptName);
While (sqlcode == 0)
{
Printf("Name of the employee: %s %s",FirstName,Surname);
Printf("Salary: %d\n",Salary);
$ fetch DeptEmp into : FirstName, :Surname, :Salary;
}
$ close cursor DeptEmp;
}
```

Certain inquiries whose outcomes are ensured to comprise of a solitary tuple, called scalar inquiries, have a more straightforward interface with the programming language. We do not have to characterize a cursor on them, yet we can utilize the into clause. Along this line, we can straightforwardly set up to which factors of the program the result of the query must be allocated. Give us a chance to inspect how the syntax of the select direction is expanded.

SelectSQL ::=select *TargetList* [into *VariableList*]
from *TableList*
[where *Condition*]
[group by *GroupingAttributeList*]
[having *AggregateCondition*]
[order by *OrderingAttributeList*]

Example:

```
$ select Firstname, Surname into :EmpFNmae, :EmpSurname
from Employee
where Number = :EmpNumber
```

The values of the traits FirstName and Surname of the employees whose enlistment number is settled in the variable EmpNumber will be replicated into the factors EmpFName and EmpSurname.

Dynamic SQL

There are numerous circumstances in which it is important to enable the client to figure subjective inquiries on the database. In the event that the queries have a predefined structure and the part that differs is only the parameter utilized as a part of the inquiry, we can develop the application by methods for cursors, as delineated in the cases in the past area. There are different cases in which it is important to oversee inquiries that require greater adaptability. These inquiries vary in the parameters utilized, as well as into the structure of the queries and in the arrangement of tables being gotten to.

The instruments for calling SQL statements found in the past area will not work in this unique circumstance, given that they require the structure of the queries to be predefined. This group of systems is by and large known as static SQL. An option group of charges allow the utilization of dynamic SQL.

These statements make it conceivable to build a program that executes SQL statements built when the program is run. These charges require unique help from the system. The best issue to be overcome is the exchange of parameters between the SQL statement and the program, both approaching and active. Given that the SQL statement is self-assertive, the program has no chance to get of perceiving at the time of arrangement, which are the required information and yield parameters of the charge. This data is fundamental for the program to have the capacity to compose the inquiry inside.

The utilization of dynamic SQL modifies the method of communication with the system. In static SQL, the charges can be handled by a compiler, which examines the structure of the charge and builds an interpretation into the inside language of the system. Along this line, orders should not be interpreted and streamlined each time, yet can be executed promptly. This brings significant focal points as far as execution. On the off chance that a summon is completed over and again, with this arrangement the interpretation is made just once, though interfacing with the motor each different execution of the charge would require its own period of interpretation.

In unique SQL, endeavours are made to hold these points of interest as much as conceivable, making accessible two distinct sorts of collaboration. The question can be done quickly, implying that the execution of the inquiry takes after quickly after the investigation, or, then again, the administration of the question occurs in two stages, investigation and execution.

Quick Execute Command

'The execute immediate' command requires the execution of a SQL explanation, either straightforwardly expressed or contained in a program variable of sort series of characters.

Execute immediate *SQLStatement*

The immediate strategy can be utilized just for articulations that don't require parameters for either information or yield. A case of the utilization of the command is shown below.

execute immediate

"delete from Department where Name = 'Administration'"

In a C-program we could compose:

SQLString =

"delete from Department where name = 'Administration'";

...

$ execute immediate :SQLString

When a statement is executed more than one time or when the program should arrange a trade of information parameters with the query, we should recognize the two stages.

Planning Stage

The 'prepare' order breaks down and advances an SQL explanation and makes an interpretation of it into the inward procedural language of the DBMS.

prepare *CommandName* from *SQLStatement*

The SQL statement can contain input parameters spoke to by a query mark. For instance:

prepare *:SQLStatement*

from "select City from Department where Name = ?"

For this situation, the interpretation of the inquiry relates to the variable SQLStatement, with a section parameter that relates to the name of the office that must be chosen by the question. At the point when an arranged SQL proclamation is never again required, we can free the memory possessed by the deciphered proclamation by utilizing the deallocate prepare command, with the following statement below.

deallocate prepare *CommandName*

Execution Stage

To command a statement expounded by set up, the execute command is utilized with the following syntax.

execute *CommandName* [into *TargetList*] [using *ParameterList*]

The objective rundown contains the rundown of parameters in which the consequence of the execution of the announcement must be composed. Here is an illustration:

execute :SQLStatement into :city using :department

Assume that the string Production is appointed to the Name in the Department. The impact of this order is to do the query found below.

select City from Department where Name = 'Production'

also, as an outcome to acquire the string Toulouse in the variable City as a final result.

Utilization of Cursors with Dynamic SQL

The utilization of cursors with dynamic SQL is fundamentally the same as the utilization made with the static SQL. However, there are two differences. The first is that the inquiry identifier is doled out to the cursor rather than to the inquiry itself. The second is that the orders for utilization of the cursor permit the detail of the into and utilizing provisions, which permit the particular of conceivable information and yield parameters. One case of the utilization of a dynamic cursor is as per the following, in which we assume that the inquiry characterized in the string SQLStatement permits an info parameter:

prepare :SQLStatement from :SQLString

declare PrgCursor cursor from :SQLStatement

open PrgCursor using :PrgVariable

Procedures

Standard SQL takes into consideration the meaning of systems, otherwise called 'stored procedures' since they are generally put away in the database as parts of the outline. Similarly as with programming languages, strategies make it conceivable to connect a name with an SQL statement, with the likelihood of determining parameters to be utilized for correspondence to the methodology. The preferences are an expansion in the clearness of the program, less demanding upkeep, and in numerous cases a recognizable increment in productivity. Once the technique has been characterized, it is dealt with as a component of the arrangement of predefined SQL charges. As a first illustration, let us see the following SQL statement, which updates the name of the city of a department.

procedure AssignCity(:Dep char(20), :City char(20)

update Department

set City = "City

where Name = :Dept;

The strategy is called by giving an incentive to the parameters. The following case demonstrates a conjuring of the system inside a C program, utilizing two factors: DeptName and CityName.

$ AssignCity(:DeptName, :CityName)

Standard SQL - 2 does not deal with the composition of complex methods, but is restricted to the meaning of systems made up of a solitary SQL command.

Numerous frameworks expel this restriction, driven by clients' prerequisites. The procedural expansions proposed by numerous systems vary generally among themselves. There are systems that permit just arrangements of orders inside a strategy as well as others that permit the utilization of control structures, announcements of neighbourhood factors, and the conjuring of outside projects. In each case, the utilization of these capacities is outside the extent of the SQL - 2 standards and renders the SQL code in this manner produced non-compact. SQL - 3 expands this part of the language and gives a rich language structure to the meaning of techniques. Meanwhile, in the event that we choose to utilize the procedural expansions given by a framework, we should likewise rework them in the event that there is an actual need to move the application to one other condition.

The accompanying case demonstrates a (non-standard) system made out of a grouping of two SQL guidelines. The technique makes it conceivable to provide for the property City the NewCity value in the Department and Employee tables in which the credit is equivalent to OldCity.

```
procedure ChangeAllCities( :NewCity char(20), :OldCity char(20))
begin
    update Department = :NewCity
    where City = :OldCity;
    update Employee
    set City = :NewCity
    where City = :OldCity;
end;
```

One of the augmentations as a rule given by the current system is the control structure which permits the declaration of contingent executions and can be utilized to deal with excellent conditions. We demonstrate a case of a system that allots to City of the Department with DeptName value in Name and NewCity value in City; if the Department is not discovered, a line is embedded into DeptEtrror.

```
procedure ChangeCity(:DeptName char(20), :NewCity char(20))
begin
    if not exists(select *
                  from Department
```

```
                    where Name = :DeptName)
          insert into DeptError values(:DeptName)
     else
          update Department
          set City = : NewCity
          where Name = DeptName;
     end if;
end;
```

As we have just demonstrated, there are business systems that offer effective procedural expansions of SQL. These augmentations are regularly ready to make the language computationally total, which is actually, to give it to the same expressive power of a conventional programming language. The likelihood exists accordingly, of composing a whole application with this expanded SQL.

This is infrequently the best arrangement on the grounds that for the most part, the relational system is advanced just for information get to. Finally, let us take a gander at a case of a program written in PL/SQL, the augmentation of the relational system Oracle Server, to give a thought of the level of usefulness offered by these frameworks. PL/SQL:

```
procedure Debit(ClientAccount char(5), Withdrawal interger) is
     OldAmount integer;
     NewAmount integer;
     Threshold integer;
begin
     select Amount, OverDraft into OldAmount, Threshold
          from BankAccount
          where AccountNo = ClientAccount
          for update of Amount;
```

```
    NewAmount : = OldAmount – Withdrawal;
    if NewAmount > OverDraft then
        update BankAccount set Amount = NewAmount
        where AccountNo = ClientAccount;
    else
        insert into OverDraftExceeded
        values(ClientAccount, Withdrawal, sysdate);
    end id;
end Debit;
```

The illustration demonstrates a method that subtracts Withdrawal from the accounts with the codes ClientAccount, if there is sufficient cash in the account. The methodology utilizes nearby factors (Old-Amount, New-Amount and also Threshold) which applies to the control structure ‘if-then-else’.

CHAPTER 5

TECHNIQUES AND MODELS OF DESIGN

In the former sections we investigated database models and languages, accepting the presence of the database with which the clients were to connect. In this section, we will start to manage the issue of planning a database based on the requirements set by the clients. Outlining a database implies characterizing its structure, attributes, and substance. As we might envision, this is a procedure in which numerous sensitive choices must be taken. The utilization of the proper methods is fundamental for the formation of a high quality item.

At this part, we will start to take a gander at the issue of database configuration when all terms and reasonable techniques for this employment are identified. Later we will outline the Entity-Relationship model that furnishes the creator with methods for information portrayal, known as the theoretical composition, which is utilized amid the main period of database plan. The outline procedure is isolated into three (3) primary stages: conceptual design, logical design, and physical design. The first two will be presented in the following sections.

THE DATABASE CONFIGURATION PROCESS

The Life Cycle of Information System

Database configuration is only one of the numerous exercises in the advancement of a information system inside an association. It ought to be introduced inside a more extensive setting of the information system life cycle. As presented in Figure 79, the life cycle of a information system for the most part comprises of the following activities.

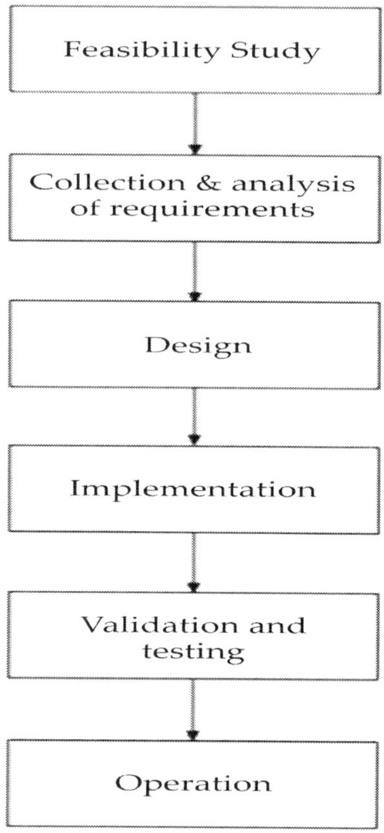

Figure 79 - Informational Systems Life-cycle.

(Source: Teorey, T. J. Design of Database Structures.)

- Feasibility Study. This serves to characterize, as exactly as would be prudent, the expenses of the different conceivable arrangements and to build up the needs in making the different parts of the framework.
- Collection and Analysis of Requirements. This comprises of the definition as well as investigation of the properties and usefulness of the information system. It requires communication with the clients so as to know the application

necessities. It delivers a total depiction of the information included and of the operations to be completed on it. The equipment and programming necessities of the information system are additionally settled.

- Design. This is partitioned into two undertakings: database plan and operational plan. In the database plan, the important structure and association of the data are built up; while in the operational plan, the qualities of the application programs are characterized. The phases are integral and can happen simultaneously or sequentially. The depictions of information and projects delivered amid this action are formal and allude to particular models.
- Implementation. This comprises of the formation of the information system as indicated by the structure and the attributes characterized in the outline movement. The database is built, populated, and the program is coded.
- Validation and Testing. This is to check the appropriateness and nature of the information system. The tests include all the required tasks or transactions.
- Operation. This is the action in which the information system is being used. Accepting there are no real blunders to be repaired or changes to make, this action requires just administration and maintenance.

It ought to be focused on that the procedure is once in a while entirely successive, given that amid one of the exercises above it is frequently important to reevaluate choices made amid the previous one, in this manner framing a ‘cycle’. In addition, infrequently another movement is included, called prototyping, which comprises of the utilization of particular programming devices for the fast formation of a rearranged variant of the data framework, with which to test its usefulness. The model can be appeared to the clients to check that the abnormal state necessities of the data framework were accurately gathered and displayed. This action is regularly the reason for the alteration of the necessities and conceivably the update of the venture.

The database constitutes just a single of the segments of a data framework, which additionally incorporates application programs, user interface, and other capabilities. Nonetheless, the focal part that the data plays in the information system legitimizes a free investigation of database outline. Therefore, we manage just those parts of data framework improvement that are nearly of databases, concentrating on information outline as well as on the related exercises of accumulation and examination of the requirements.

This procedure is with regards to the information driven way to deal with information system advancement, in which consideration is focused on the information and its properties. With this approach, the database is composed to begin with, trailed by the applications that utilize it.

Methodologies for Database Framework

In this part, we take after an organized way to deal with database plan that can be viewed as a 'plan procedure'. Thus, it is introduced by methods for:

- a deterioration of the whole outline action in progressive strides, autonomous one from the other;
- a progression of procedures to be followed in the different strides and a few criteria from which to pick on account of there being choices; and
- some reference models to depict the information sources and yields of the different stages.

The properties that such an approach must ensure include:

- consensus as to the application and the systems which are in play (and along these lines the likelihood of utilization that is autonomous of the particular application and of the accessible systems);
- the item quality regarding precision, fulfillment, and effectiveness; and
- the usability both of the techniques and of the reference models.

Inside the field of databases, a planning has been merged throughout the years, which fulfills every one of the properties depicted. This procedure is partitioned into three stages to be taken after successively (see Figure 80). It depends on a basic but exceptionally effective building guideline: that of neatly isolating the choices identifying with "what" to speak to in the database (first stage), from those identifying with "how" to do it (second and third stages).

- Conceptual plan. The reason for this is to speak to the casual necessities of an application as far as a formal and finish portrayal, yet free of the criteria for portrayal utilized as a part of database administration system. The result of this stage is known as the calculated pattern and alludes to an applied information demonstrate. Calculated models permit the depiction of the association of information at an abnormal state of reflection, without taking into account the execution viewpoints. In this stage, the creator must attempt to speak to the data substance of the actual database, without considering either one of the methods by which the data will be executed in the genuine system, or the effectiveness of the projects that utilize these data.

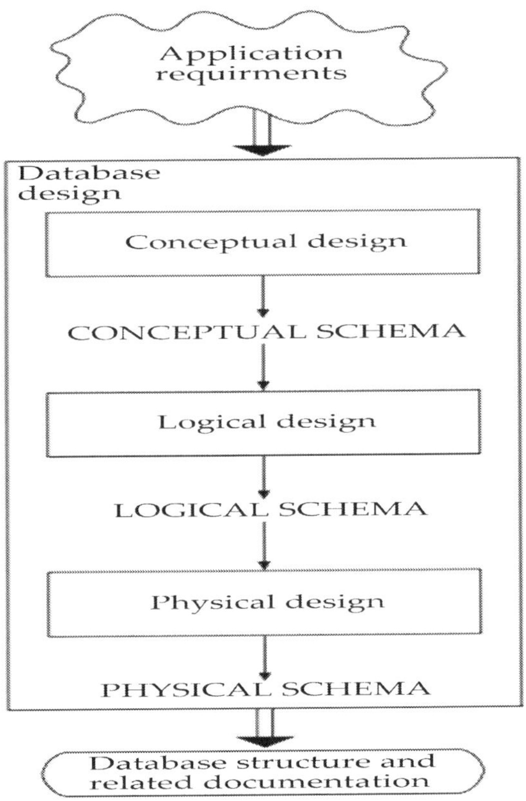

Figure 80 - Database Design Stages.

(Source: Teorey, T. J. Design of Database Structures.)

- Logical outline. This comprises of the interpretation of the calculated diagram characterized in the previous stage, into the information demonstrate received by the database administration system. The result of this stage is called the sensible outline of the database and alludes to a coherent information demonstrate. As we have noticed, a sensible model speaks to information in a way that is still autonomous of the physical points of interest, in spite of the fact that the DBMS utilized for the execution must be one that backings that information show. In this stage, the architect should likewise consider some streamlining criteria based on the operations to be completed on the information. Formal strategies for confirmation of the nature of the intelligent pattern are frequently utilized. For the situation of the relational data model, the most generally utilized system is that of standardization.
- Physical plan. In this stage, the intelligent pattern is finished with the points of interest of the physical execution (record association and lists) on a given DBMS. The result of this stage is known as the physical model and alludes to a physical data model. This model relies upon the particular database administration framework picked and considers the criteria for the physical association of the data in that framework.

Let us see how the application requirements are utilized as include in the plan. We can recognize information necessities, concerning the substance of the database, operational necessities, and utilization of the database by clients or projects. In applied plan, information necessities give a large portion of the data, though operational necessities are utilized just to confirm that the applied diagram is finished (It contains the data important to complete all the operations that will be required). In consistent plan, the theoretical framework, given as information, abridges the information necessities. Through the operational requirements with the anticipated application stack are utilized to get a sensible diagram, which takes into consideration the effective execution of such operations. Lastly, in the physical model, the coherent mapping and the operational are utilized to advance the execution of the data framework. In this stage, it is important to take into account the qualities of the identified DBMS.

The aftereffect of the outline procedure of a database is not only the physical mapping but also the theoretical pattern and the consistent outline. The applied blueprint gives an abnormal state portrayal of the database, which can be exceptionally helpful for documentation purposes. The sensible construction gives a portrayal of the substance of the database that, leaving aside the execution angles, is valuable as a source of perspective for composing queries and updates (Date C. J.).

In Figure 81, we demonstrate the results of the different stages on account of the plan of a relational database in light of the utilization of the best known data model, which is the Entity-Relationship model. An Entity-Relationship model presented as a chart depicts the database at reasonable level. This portrayal is then converted into a relational table, made up of a many tables. Lastly, the information is depicted from a physical perspective (sort and size of the fields) and helper structures, like records, are determined for effective access to information.

In the following sections we will inspect in detail the different exercises of database configuration portrayed in Figure 80 and with reference to the models utilized as part of Figure 81. The Entity-Relationship model is perceived as the standard data model to illustrate a database design. Calculated outline, which we will talk about in the following part, depends on this model. Amid sensible outline we will utilize the relational data model, which as we have seen in before sections, is the most generally utilized by current DBMS.

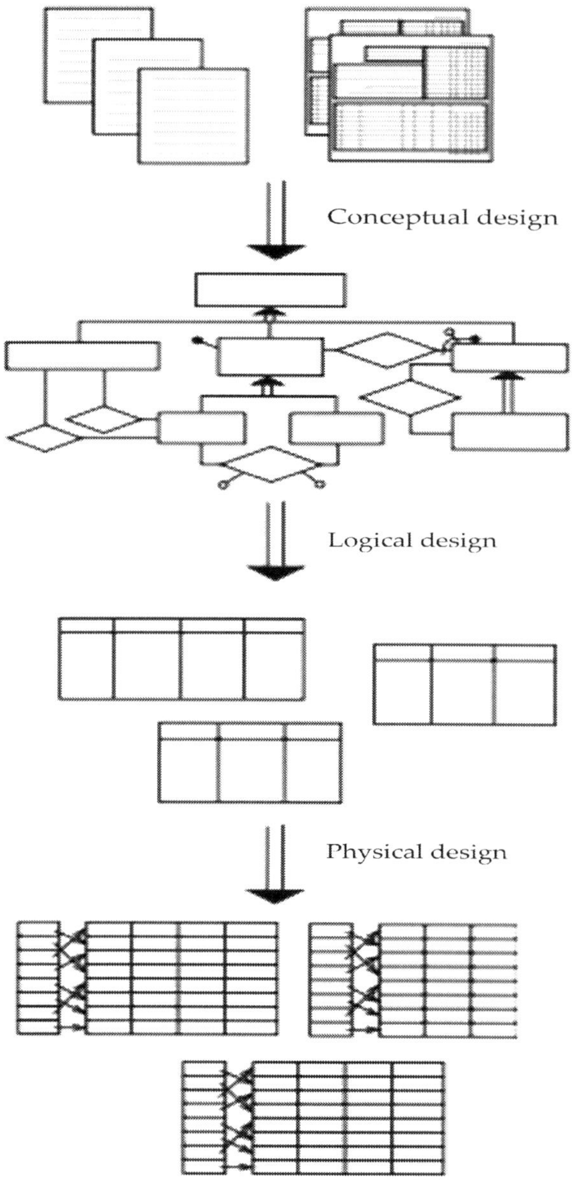

Figure 81 - The Products, Various Phases of the Design with the E-R Model of a Relational Database.

(Source: Teorey, T. J. Design of Database Structures.)

THE ENTITY-RELATIONSHIP DIAGRAM

The Entity-Relationship (E - R) model's information display gives a progression of builds equipped for portraying the information necessities of an application in the way that it is straightforward and is free of the criteria for the administration and association of information on the system. These builds are utilized to characterize mappings, which portray the association and direct the legitimate occurrences of information, are the real esteems that the database could hold at various circumstances.

In the table shown in Figure 82, the development of an E-R model requires corresponding symbols. As we might see, this portrayal enables us to characterize a composition diagrammatically.

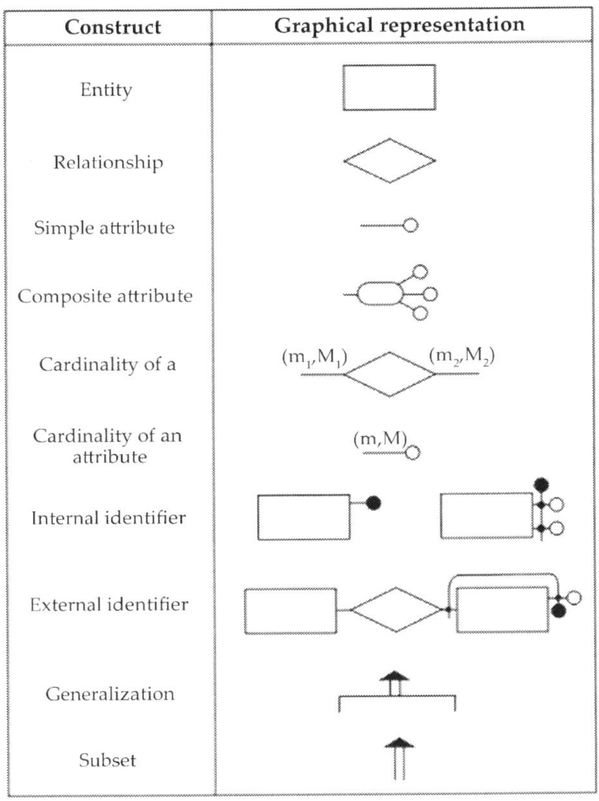

Figure 82 - The Constructs of the E-R Model.

(Source: Teorey, T. J. Design of Database Structures.)

The Essential Parts of the Model

We will start by breaking down the principle parts of this model: entities, relationships, and attributes.

Entities

These refer to classes of items (actualities, things, individuals) that have similar properties and are named in singular form: CITY, DEPARTMENT, EMPLOYEE, PURCHASE, and SALE are examples of an entity in an application for a business association. An instance of an entity is a protest of the class that the element speaks to. The urban communities of Stockholm, Helsinki, and Oslo are instances of the CITY entity, while the workers Petersen and Johanssen are instances of the Worker entity. Note that an instance of an entity is not a value that recognizes a query (such as the surname of the worker or a relational security number) yet it is simply the query (the worker 'in the tissue'). An intriguing result of this reality is that an instance of an entity has a presence (and a personality) autonomous of the properties related with it (On account of an employee, he or she exists paying little heed to having name, surname, age, and so forth). In this regard, the E-R model demonstrates a checked distinction from the social model in which it is impractical to speak to a protest without knowing its properties. (A worker is spoken to by a tuple containing the name, surname, age, and different traits.)

In a construction, each entity has a unique name and is graphically represented by a rectangle containing the entity name. Figure 83 shows examples of entities.

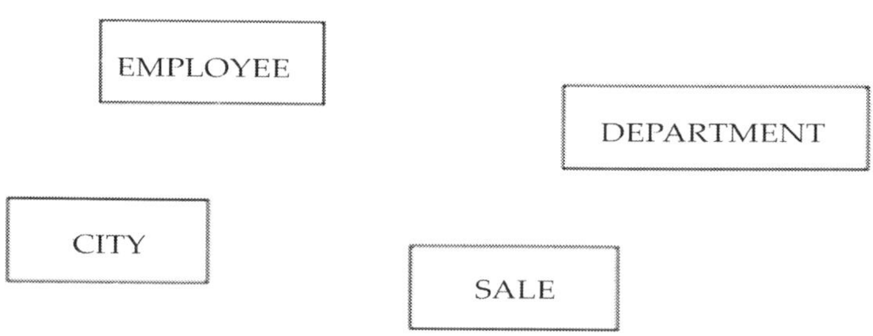

Figure 83 - Entities of E-R Model.

(Source: Teorey, T. J. Design of Database Structures.)

Relationships

These refer to consistent connections between at least two entities. Habitation is a relationship that can exist between the CITY and EMPLOYEE entities, while EXAM is a relationship that can exist between the STUDENT and COURSE entities. An instance of a relationship is an n-tuple (a couple in the most continuous instance of the double relationship) made up of instances of intities, one for each of the entities included. The match of items made up of the employee who has a name of Johanssens and the city named Stockholm, or the combination of items produced using the employee with a name of Petersen and the city which is named Oslo, are instances in the relationship RESIDENCE. Instances of the relationship EXAM between STUDENT and COURSE entities are the sets e1, e2, e3, e4, e5, and e6 appeared in Figure 84, in which the instances of the entities included are additionally appeared (Fairly, 1985).

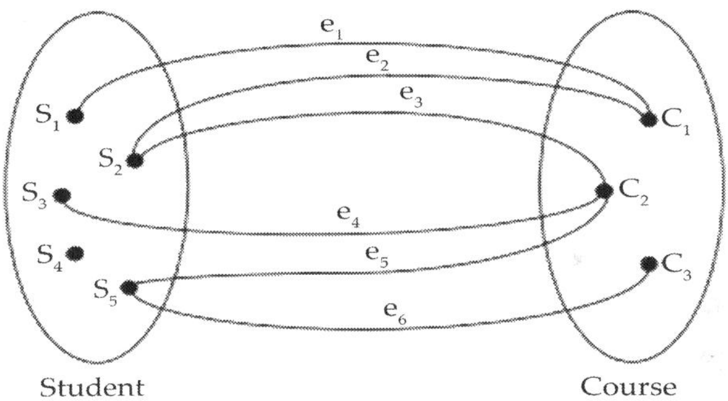

Figure 84 - Exam Relationship.

(Source: Teorey, T. J. Design of Database Structures.)

In the E-R model, every relationship has an extraordinary name and is graphically represented by a diamond containing the name of the relationship and with lines that associate the diamond with every entity in the relationship. Figure 84 indicates cases of constructions with connections between entities. Note that there can be distinctive connections between the same entities, for example, the WORKPLACE and RESIDENCE relationship between the EMPLOYEE and CITY entities. In naming the relationship, it is desirable to use noun over verb, so as not to infer or recommend an action in the relationship. For instance, WORKPLACE is desirable over WORKSIN.

An imperative part of connections is that the arrangement of an instance of relationship is, to all goals and purposes, a numerical connection between the events of the included substances, as appeared by Figure 85. A numerical connection on two arrangements of entities is a subset of their Cartesian item. This implies no n-tuples can be rehashed among the instances of a relationship of the E-R model. This perspective has imperative outcomes: for instance, the relationship EXAM in Figure 85 does not have the limit to report the way that a specific student has taken a similar exam more than once (in light of the fact that this would deliver indistinguishable sets). For this situation, the exam ought to be spoken to by an entity alone, connected to the STUDENT entity as well as the COURSE entity by means of twofold connections.

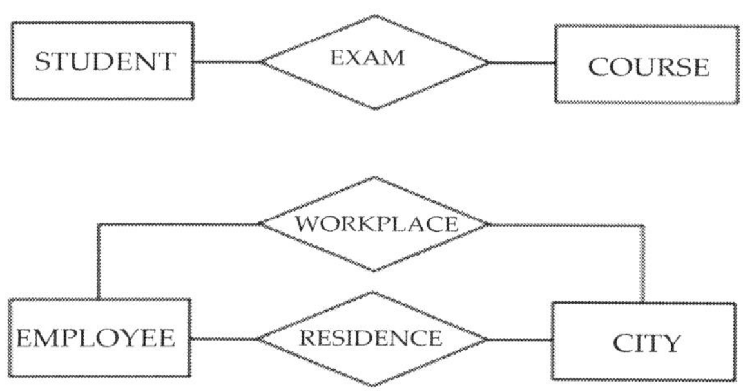

Figure 85 - Relationship Examples.

(Source: Senn, J. A. Analysis & Design of Information Systems)

Recursive connections are additionally conceivable, that is connections between an entity and itself. For instance, in Figure 86, the recursive relationship COLLEAGUE on the EMPLOYEE entity associates sets of individuals who work together, while the relationship SUCCESSION on the SOVEREIGN entity designates the following in line to every sovereign of an administration. Note that, dissimilar to the first relationship, the relationship SUCCESSION is not symmetrical. For this situation it is important to show the two parts that the substance included plays in the relationship. This can be accomplished by identifiers with the lines exuding from the recursive relationship (for this situation, Successors and Predecessors).

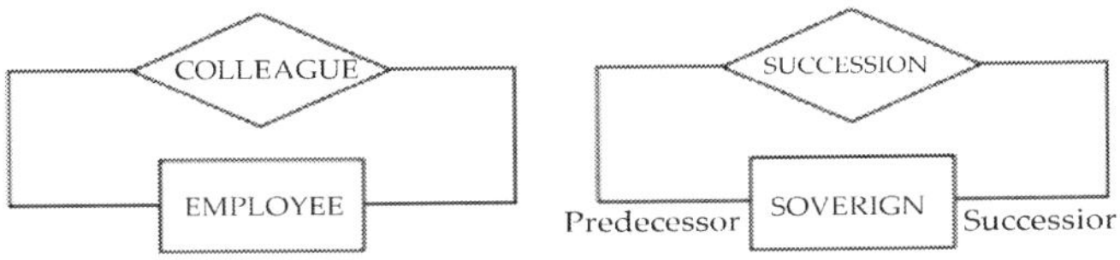

Figure 86 - Recursive Relationship.

(Source: Senn, J. A. Analysis & Design of Information Systems)

Finally, we can have connections that include more than two entities. A case is presented in Figure 87: an event of the relationship SUPPLY portrays the way that a given SUPPLIER supplies a specific PRODUCT to a DEPARTMENT.

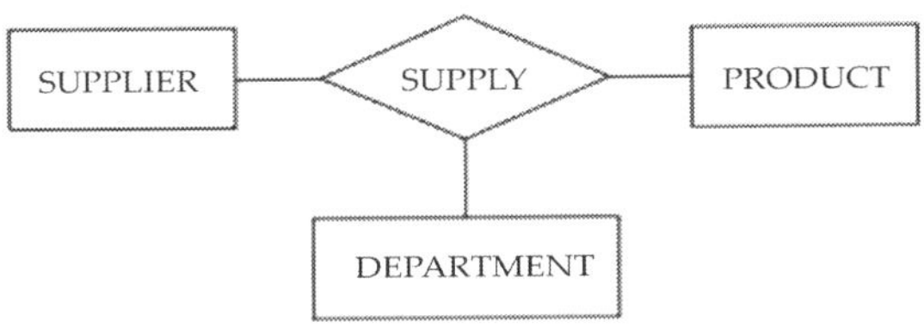

Figure 87 - Ternary Relationship.

(Source: Senn, J. A. Analysis & Design of Information Systems)

A conceivable arrangement of instances of this relationship could build up that the Acme firm supplies printers to the business office and adding machines to the investigate division, while the Nakio firm supplies printers to the deals office. A graphical portrayal of the conceivable instances of the SUPPLY relationship is illustrated in Figure 88 (triples s1, s2, s3, and s4). In the outline, the instances of the entities included are additionally appeared.

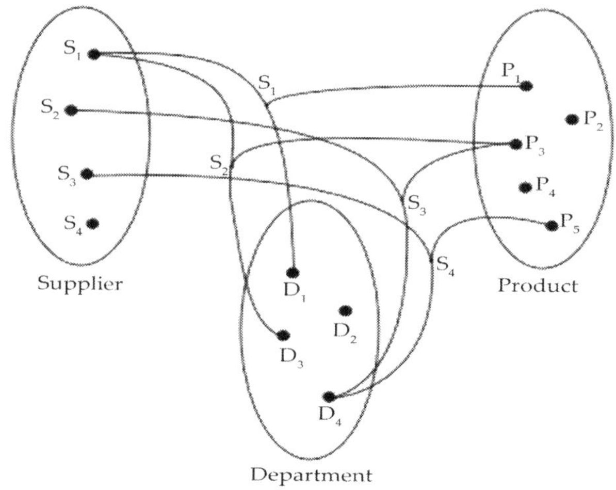

Figure 88 - Supply Relationship.

(Source: Senn, J. A. Analysis & Design of Information Systems)

Attributes

These depict the basic properties of entities or relationships. For instance, Surname, Salary, and Age are conceivable attributes of the EMPLOYEE entity, while Date and Mark are conceivable elements for the relationship EXAM between STUDENT and COURSE. An attribute partners with every instances of an entity (or relationship) with esteem in a set known as the space of the characteristic. The area contains the allowable values for the characteristic. For instance, the characteristic Surname of the Worker entity can have any of the 20-character string as a space, while the Age attribute can have as a space in numbers from 18 to 65. Figure 5.89 demonstrates how the characteristics are spoken to graphically. Areas are definitely not appeared, as they are typically portrayed in the related documentation.

It can be helpful to gather attributes of a similar entity or relationship that have firmly associated implications or employments. The arrangement of attributes got in this way is known as a composite property. We can, for illustration, amass together the attributes Street, HouseNumber, and PostCode in the PERSON entity to frame the composite attribute Address. A graphical portrayal of a composite property is appeared in

Figure 90. Composite attributes can be exceptionally valuable as methods for condensing a few related qualities, for instance, every single clinical exam of a patient, precluding their nitty gritty portrayal.

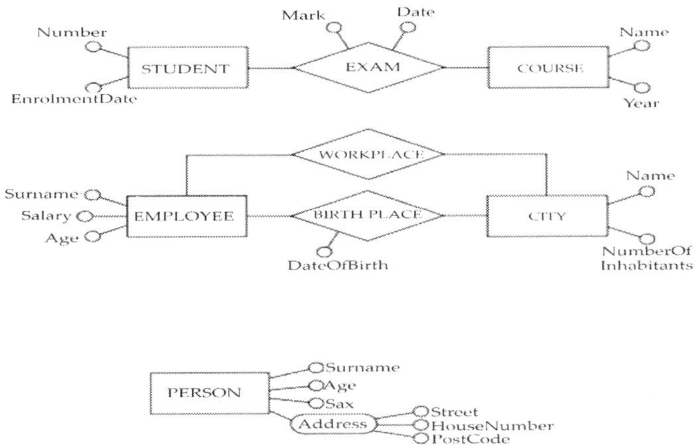

Figure 89 - E-R Schemas.

(Source: Senn, J. A. Analysis & Design of Information Systems)

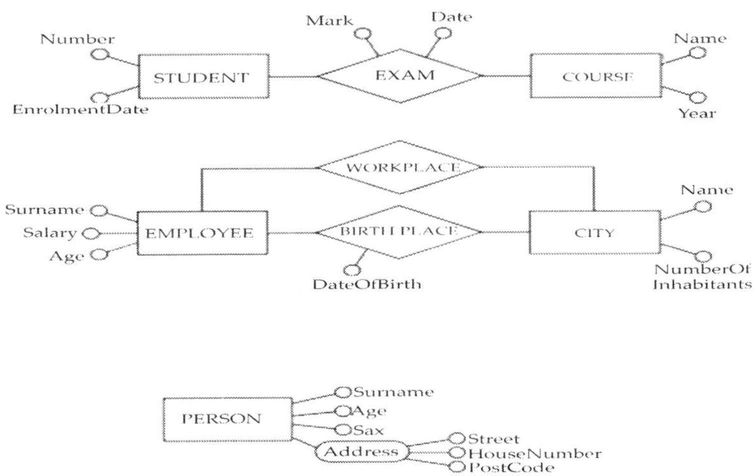

Figure 90 - *Entity Example with a Composite Attribute.*

(Source: Senn, J. A. Analysis & Design of Information Systems)

Other Parts of the Model

Give us now a chance to look at the other parts of the E-R model: the cardinality of relationships and attributes, the identifiers and the speculations. As we might see, just the latter is "another" part. The others constitute uprightness limitations on parts we have just observed, attributes that must be fulfilled by the instances of entities and relationships with a specific end goal to be viewed as 'substantial'.

Cardinality of Relationships

These are determined for every entity taking an interest in a relationship and depict the most extreme and least number of relationship instance in which an instance can take an interest. They state along these lines, how often in a connection between instances of an entity can be connected to instances of alternate entity included. For instance, assume that in the ASSIGNMENT relationship between the EMPLOYEE and TASK entities we determine for the first entity a base cardinality 1 and a most extreme cardinality 5. This implies that an employee one to five instances of the ASSIGNMENT relationship. In our application, an employee has an assignment of one to five tasks.

Further, assuming that the TASK entity we determine a base cardinality of equivalent to 0 and a greatest cardinality equivalent to 50. For this situation, it shows that a task has none to fifty instances of the ASSIGNMENT relationship. Hence, a task has none to fifty employees assigned to it. Figure 91 shows the E-R model cardinality being described above.

Figure 91 - E-R Model Cardinality.

(Source: Batini, C. C. Conceptual Database Design, an Entity-Relationship Approach.)

On a basic level, it is conceivable to allocate any non-negative number to the cardinality of an association with the base cardinality lesser or equal to most extreme cardinality. As a rule, it is adequate to utilize just three values: 0, 1, and the letter "N" (which is called "numerous" and represent a whole number greater than 1). Specifically:

- for the base cardinality, 0 or 1; in the primary case we say that the support in the relationship is discretionary, in the second we say that the support is obligatory; and
- for the greatest cardinality, one or numerous (N); in the first case each event of the element is related at most with a solitary event of the relationship, while in the second case every event of the substance is related with a subjective number of events of the relationship.

Give us a chance to re-examine Figure 84. It demonstrates that the STUDENT entity takes an interest in the EXAM association with a cardinality equivalent to (0,N). This implies there are students who do not take an interest in any event of the relationship (S4), while other students take an interest in more than one event of the relationship (for instance, S2 who takes an interest in e2 and e3).

In Figure 92, different instances of cardinality for connections are appeared. For illustration, the cardinality of the RESIDENCE relationship reveals that each person is a resident to one and only one city, while every city has none to many persons residing in it.

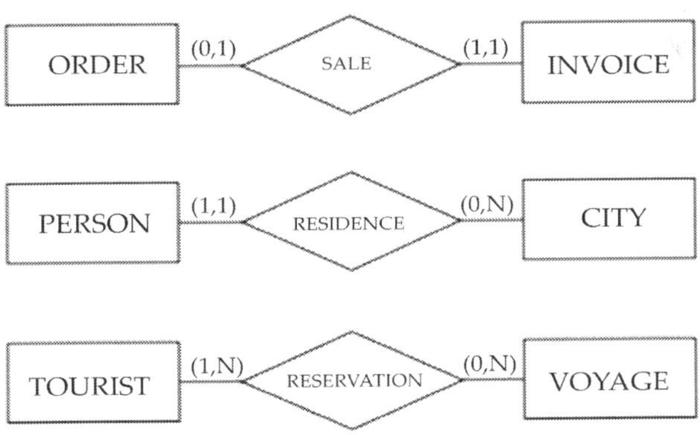

Figure 92 - Cardinality Relationships.

(Source: Batini, C. C. Conceptual Database Design, an Entity-Relationship Approach.)

In Figure 92, characterize a coordinated correspondence between such elements and are in this manner called balanced connections. Likewise, connections between an element with most extreme cardinality equivalent to one and another with most extreme cardinality equivalent to N, for example, the relationship Habitation in Figure 92, are called one-to-many connections. Finally, connections having a most extreme cardinality equivalent to N for both the substances included, like the relationship RESERVATION in Figure 92, are called many-to-many connections. For least cardinality, take note of that the instance of compulsory investment for every one of the elements included is uncommon. This is on account that when another instance is included; all the time the relating instances of alternate entities connected to it are not yet known or even do not exist. For instance, consider the model in Figure 92. At the point when another arrange is gotten, there exists so far no receipt identifying with it and it is unrealistic to build an instance for the SALE relationship that contains the new request (Senn, 1989).

In n-ary connections, the entities included quite often have a greatest cardinality N. An illustration is given by the ternary relationship SUPPLY in Figure 87. As should in Figure 88, there are cases of events of each entity (S1, P3, and D4) that show up in numerous events of this relationship. At the point when a substance is included in a n-ary association with a most extreme cardinality equivalent to one, it implies that one of its events can be connected to a solitary event of the relationship, and consequently to a solitary n-tuple of events of other entity associated with the relationship. This implies it is conceivable (and on occasion can appear to be more normal) to supplant the n-ary association with n twofold one-to-many relationships that connects such an entity with the others.

Figure 93 demonstrates a case of an entity which attributes traits have cardinality. Because of cardinalities, we realize that a person has one and just a single surname, can have a driving permit (however in the event that he or she has one, it is one of a kind) and can have none or few autos.

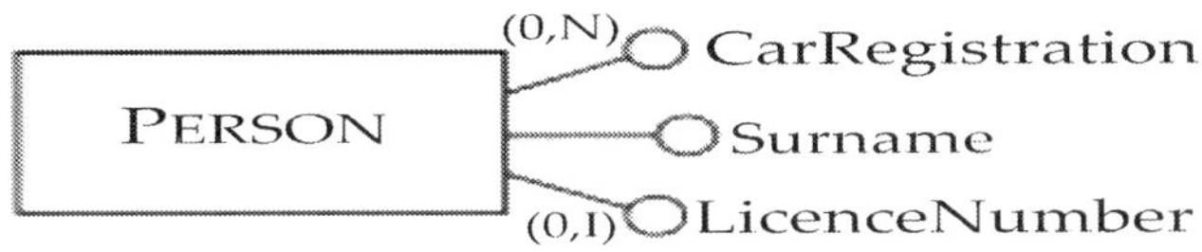

Figure 93 - Entity Attributes with the Cardinality.

(Source: Batini, C. C. Conceptual Database Design, an Entity-Relationship Approach.)

In view of what we have said regarding the matter of identifiers, we can make a few general perceptions:

- an identifier can include at least one attribute, given that each of them has (1,1) cardinality;
- an outside identifier can include at least one elements, given that each of them is individual from a relationship to which the entity to distinguish takes an interest with cardinality equivalent to (1,1);
- an outer identifier can include a substance that is in its turn distinguished remotely, insofar as cycles are not produced; and
- every substance must have one (interior or outer) identifier, but can have more than one. All things considered, if there is more than 1 actual identifier, traits and substances engaged with a recognizable proof can be discretionary (least cardinality equivalent to 0).

Final Outline of the E-R Model

We have perceived how the Entity-Relationship model gives the originators with a few builds for portraying the information engaged with an application, related with a graphical portrayal that is straightforward.

Every one of the builds of the E - R model is delineated in the outline in Figure 94, which in the meantime gives a case of an E-R model and a streamlined depiction of the E-R model itself. Give us now a chance to break down this pattern and consider this investigation as an activity of 'perusing an E - R model'. This is an action that we have to rehearse, as it is one that happens every now and in the investigation and upkeep of existing information systems.

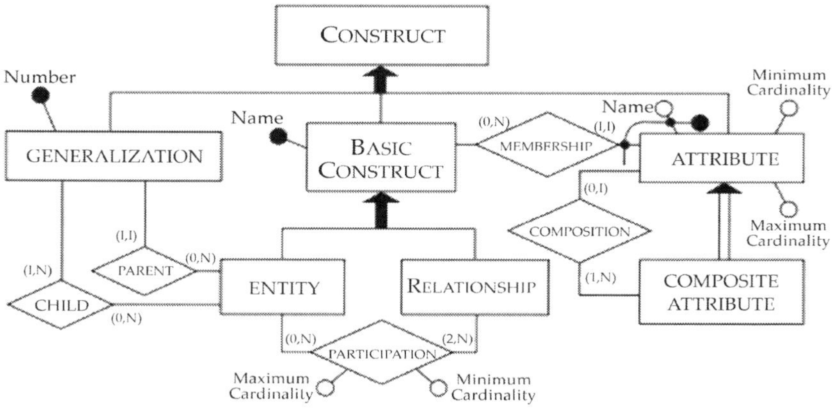

Figure 94 - E-R Model Description.

(Source: Batini, C. C. Conceptual Database Design, an Entity-Relationship Approach.)

We can actually see that the real model is comprised of a progression of develops of which two are viewed as 'fundamental': the entity and the relationship. An entity can take an interest in at least 0 relationships, while a relationship includes two or more entities. The support of an entity in a relationship has a minimum and a maximum cardinality. Other develops of the model are properties and speculations. An attribute has a name, a base and greatest cardinality, and has a place with a fundamental idea, that is to an entity or a relationship. The MEMBERSHIP relationship is acquired by the offspring of the BASIC element. Composite properties are a specialization of attributes and are comprised of at least one of the attributes.

As indicated by the cardinalities, a speculation has precisely 1 parent entity and one (on account of subsets) or more offspring. An entity can be parent or an offspring of numerous speculations (as well as of none). Lastly, a fundamental model is distinguished unambiguously by its name (it is basic in an outline to abstain from utilizing a similar name for various entities or relationships).

Two attributes can have a similar name on the off chance that they are in different connections or entities as portrayed by the outside ID. Speculations do not often have a name and to recognize them we accept that they are numbered. There exist, at last, different requirements on the utilization of models that cannot be communicated in the blueprint. For instance, a pecking order of speculations cannot contain cycles. Additionally, a base cardinality must be at least equal to the most extreme cardinality. The issue of documentation of limitations that cannot be communicated in the E-R model will be talked about completely in the following segment.

We will finish up the segment with a thought of a general sort. E-R model gives a conceptual portrayal of the information engaged with an application and they can be utilized for purposes other than database plan. There are different cases of the conceivable utilization of calculated outlines aside from plan.

- E-R models can be utilized for documentation purposes, as non-IT professionals may comprehend them.
- They can be utilized to portray the information of an information system available at present (for instance, to incorporate it with different databases). On account of the system made up of heterogeneous subsystems, there is the preferred standpoint that the E-R diagram can speak to the different parts with a unique language, which is in this manner bringing together.
- For the situation of change of the necessities of an application, E-R diagrams can be utilized to comprehend which segment of the system should be changed and what alterations are to be completed.

DOCUMENTATION OF E-R DIAGRAM

We have perceived how the Entity-Relationship diagram gives very expressive demonstrating apparatuses that permit the simple and effective portrayal of exceptionally perplexing circumstances. An E-R model is seldom adequate without anyone else's input to speak to every one of the parts of an application in detail. Right off the bat, in an E- R diagram, just the names of the different ideas show up, and this can be inadequate to clarify their importance.

Also, when a pattern is especially perplexing, it may not generally be conceivable to speak to every one of the properties of the different ideas, in a way that is fathomable. All in all, speaking to substantial compositions graphically requires a specific capacity in conceiving a decipherable format.

Different impediments concern the expressive energy of the E-R model. It is as a matter of fact unthinkable in specific cases to speak to a few properties of information by methods for the model that the E-R presents. The example in the figure assumes that in our organization an employee can be a head just of the department to which he has a place. This property cannot be straightforwardly communicated in the diagram since it alludes to two free ideas (administration and participation) depicted by two connections, and there are no develops in the model that permit two connections. Another case of properties that cannot be straightforwardly communicated by the model of the E-R diagram is the way that an employee cannot have a salary higher than the head of his or her department. Both the aforesaid properties relate to respectability requirements on the information. Truth be told, while the E-R model is adequately expressive to speak to information, it does not give appropriate intends to the portrayal of complex limitations on information (O'Neil P. , 1994).

Accordingly, it is essential to give each E-R model with help documentation, which can encourage the translation of the mapping itself as well as portray properties of the information that cannot be communicated specifically by the parts of the model.

The Business Rules

Business rules are one of the tools utilized by data frameworks experts to depict the properties of an application that cannot be specifically communicated with an applied model. As proposed by their name, they permit the detail of "rules" of the specific application space that we are considering. The way that a worker cannot earn more than his or her head is a rule of the business.

The term business rule is regularly utilized with a more extensive importance by experts, to demonstrate any data that characterizes or obliges some part of an application. Specifically a business rule can be:

- the portrayal of an idea significant to the application or the exact meaning of an entity, an attribute, or a relationship of the E-R model;
- an uprightness imperative on the information of the application, regardless whether it be the documentation of a requirement communicated by methods of the E-R model (for instance, the cardinality of a relationship) or the portrayal of a limitation that is not specifically expressible utilizing the parts of the model;
- an induction or an idea that can be taken by induction, an arithmetic computation, or different ideas of the construction (for instance a trait Cost whose esteem can be gotten from the total of the qualities Net and Taxes).

For rules of the primary sort, it is obviously difficult to characterize an exact language structure. In addition, as a rule we need to utilize sentences in regular language. As we will depict in the accompanying area, these standards are regularly spoken to in the type of glossaries, gathering the depictions as suitable (for instance, by entity or relationship).

On the other hand, the rules that portray honesty limitations and deductions loan themselves more to the formal definitions and rather complex syntax has been proposed for communicating them. Notwithstanding, given that there are no norms and that each formalization risks not being adequately expressive, we will keep on using definitions in common languages to structure such definitions in the proper way.

Specifically, honesty requirements can simply be communicated as declarations, that is, explanations that must dependably be fulfilled in our database.

For reasons of clearness and simplicity of creation, such explanations must be 'nuclear'. That is, they cannot be deteriorated into phrases that constitute statements in themselves. Besides, since they are utilized to record an E-R pattern, attestations ought to be expressed in a revelatory way, which ought not to recommend approaches to fulfill them, in light of the fact that such proposal would give bits of knowledge on the usage, yet this is not significant to applied portrayal.

Therefore, a documentation of the sort if <condition> then <action> is definitely not reasonable to express business rules, when they record an E-R model.

Documentation Methods

We have said that an E-R model ought to be provided with help documentation, to encourage the elucidation of the diagram itself and to portray properties of the information that cannot be communicated straightforwardly by the builds of the model. We have seen that this documentation can be communicated regarding business rules. Give us now a chance to find in which shape it is conceivable to deliver this documentation, alluding to a case.

The documentation of the different ideas spoke to in a diagram, the business guidelines of the elucidating sort, can be effortlessly composed as an information word reference. This is comprised of two tables: the principal depicts the substances of the pattern with their names, a casual definition in normal language, the rundown of the considerable number of qualities (with conceivable portrayals related with them), and the conceivable identifiers. The other table portrays the associations with their names, a casual portrayal, the rundown of characteristics (with conceivable portrayals) and the rundown of the elements included, together with their cardinalities. Note how the lexicon can likewise be used to archive a few imperatives on the information and in this manner different types of business rules. As we have just shown, the utilization of the information word reference is especially vital when the construction is mind boggling and it is difficult to determine every one of the traits of element and relationship straightforwardly on the model (Lum, 1979).

In respect to other business rules, we can turn to another table, which records the different principles, regulated by sort. Such guidelines can be communicated in the shapes recommended in the area above, if conceivable alluding unequivocally to the ideas of the pattern. Keep in mind that it is essential to speak to all the decides that depict limitations not communicated in the outline, but it can once in a while be helpful likewise to speak to decides that archive limitations effectively communicated in the composition. A case of documentation of this sort is shown beneath:

Entity	**Description**	**Attributes**	**Identifier**
EMPLOYEE	Employee working in the company.	Code, Surname, Salary, Age	Code

PROJECT	Company project on which employees are working.	Name, Budget, RelcaseDate	Name
DEPARTMENT	Department of a company branch.	Phone, Name	Name, BRANCH
BRANCH	Company branch in a particular city.	City, Address (Number, Street and PostCode)	City

Relationship	**Description**	**Entities involved**	**Attributes**
MANAGEMENT	Associates a manager with a department.	Employee (0.1) Department (1.1)	
MEMBERSHIP	Associates an employee with a department.	Employee (0.1) Department (1 .N)	StartDate
PARTICIPATION	Associates employees with projects.	Employee (0.N) Project (1.N)	StartDate
COMPOSITION	Associates a department with a branch.	Department (1.1) Branch (1N)	

Figure 95 - Data Dictionary Example.

(Source: Atzeni, P. D. Relational Database Theory.)

Constraints	
(BR1)	The manager of a department must belong to that department.
(BR2)	An employee must not have a salary greater than that of the manager of the department to which he or she belongs.
(BR3)	A department of the Rome branch must be managed by an employee with more than 10 years' employment with the company.
(BRI4)	An employee who does not belong to a particular department must not participate in any project.
Derivations	
(BR5)	The budget for a project is obtained by multiplying the sum of the salaries of the employees who are working on it by 3.

Figure 96 - Rules of Business.

(Source: Atzeni, P. D. Relational Database Theory.)

CHAPTER 6

CONCEPTUAL DESIGN

The theoretical outline of a database comprises of the development of an Entity-Relationship diagram, giving an ideal depiction of the client's requirements. Notwithstanding for basic applications, the diagram can contain numerous ideas, corresponded in a very unpredictable manner. It takes after that the development of the last pattern is, frequently, an incremental procedure. The applied pattern is refined and enhanced amid a progression of changes and conceivably revisions. In this section, we portray the techniques that can be taken after the improvement of an applied pattern.

Before we start to examine this technique, it merits committing a few regard for the movement that goes before the real plan process itself: the gathering and examination of the requirements. This stage is not so much isolated from plan, rather frequently covered. The development of an E-R model starts a long time before the accumulation of requirements is finished and it can impact advance accumulation action. The statement of requirements and investigation is trailed by an introduction of some broad criteria for changing casual necessities into Entity-Relationship parts. We at that point, proceed onward to show the most imperative plan systems and to break down the qualities that a very much composed reasonable pattern must have.

REQUIREMENTS GATHERING AND EXAMINATION

It should first be expressed that the exercises of requirements gathering and examination are hard to institutionalize, on the grounds that they depend enormously on the application. In any case, we will examine some functional guidelines that it is useful to follow in this period of the improvement of a database.

By requirements gathering, we mean the total ID of the issues that the application must illuminate and the components that ought to describe such type of an application. By this, we mean both of the static perspectives (the information) and the dynamic perspectives (the operations on the information). The requirements are first accumulated into determinations that are by and large communicated in normal language. In this way, they are regularly questionable and complicated. The requirements investigation comprises of the illumination and association of the particular requirements. The requirements originate from various sources, such as the following.

- The clients of the application. For this situation, the data is obtained through meetings or by methods for archives particularly composed and traded for this reason.
- All the current documentation that has some association with the issue: shapes, inward guidelines, business methodology, laws and controls, and so forth. For this situation, gathering and choice are required. The client will normally help here, however the obligation rests with the originator.
- Possible prior applications that are to be supplanted or that must connect in some path with the new application. The learning of these product bundles (record groups, screen shapes, calculations, and related documentation) can furnish us with vital data that moreover identifies with existing issues that must be settled.

We should focus on the fact that, in the requirements gathering process, a vital part is played by the cooperation with the clients. Amid this association, the different clients can give unique data, frequently correlative, yet sometimes they do conflict. In general, the more clients involved, the more extensive but less definite.

Training Company
We wish to create a database for a company that runs training courses. For this, we must store data about the trainees and the instructors. For each course participant (about 5000), identified by a code, we want to store the social security number, surname, age, sex, place of birth, employer's name, address and telephone number, previous employers (and period employed), the courses attended (there are about zoo courses) and the final assessment of each course. We need also to represent the seminars that each participant is attending at present and, for each day, the places and times the classes are held. Each course has a code and a title and any course can be given any number of times. Each time a particular course is given, we will call it an 'edition' of the course. For each edition, we represent the start date, the end date, and the number of participants. If a trainee is a self-employed professional, we need to know his or her area of expertise, and, if appropriate, his or her title. For somebody who works for a company, we store the level and position held. For each instructor (about 300), we will show the surname, age, place of birth, the edition of the course taught, those taught in the past and the courses that the tutor is qualified to teach. All the instructors' telephone numbers are also stored. An instructor can be permanently employed by the training company or can be freelance.

Figure 97 - Natural Language Example.

(Source: Atzeni, P. D. Relational Database Theory.)

They can guide the fashioner to the specialists on person issues. As we said before, the requirements gathering instrument is frequently composed in regular languages, in any event in the main draft. Characteristic language is, by nature, subject to equivocalness and distortion. Along these lines, we have to complete an inside and out examination of the instrument so as to eliminate any mistakes and uncertain terms. To build up some down to earth rules for doing this undertaking, consider this case. Envision that we have to plan a database for a preparation organization and that we have assembled the information determinations appeared in Figure 97. The information was gathered through interviews with the organization personnel. Note that we have additionally collected data from the existing files. It is clear that such material contains various ambiguities and mistakes. For instance, we have tradable utilization of members or learners, guides or educators, and courses or classes. We will now set up a few rules for composing requirement details clearly and unambiguous (Batini, Conceptual Database Design, an EntityRelationship Relationship Approach,, 1992).

- Choose the proper level of reflection. It is savvy to keep away from terms that are excessively broad or excessively particular since they can make an idea less clear. For our situation, the terms period employed (on line-6), title (line-15), and assessment (line-7) could be determined more correctly (for instance, begin date and end date, professional title and marks out of ten).
- Standardize sentence structure. In determining requirements, it is best to utilize a similar style. For instance, 'for <concept> we hold <properties>'.
- Avoid complex expressions. Definitions ought to be kept clear and basic. Representatives are desirable over somebody who works for a company (line-15).
- Identify equivalent words and homonyms, and institutionalize terms. Equivalent words are words that have a similar meaning such as instructor (line-16) and tutor (line-18), course participant (line-3) and trainees (line-2). Homonyms are the words with more than one meaning like in place of birth (line-4) and places and times the classes are held (line-9). These circumstances can cause ambiguities and must be elucidated: for equivalent words, single terms must be picked, and for homonyms, unique terms must be utilized.
- Make cross-references express. The nonappearance of a reference among the terms could make some ideas uncertain. For instance, in line-5, it is definitely not clear whether the address and phone number are alluding to the students or to their bosses. Besides, in the expression somebody who works for … (line-15), we should illuminate to whom precisely we are alluding (trainee or instructor) keeping in mind the end goal to maintain a strategic distance from disarray.

- Construct the glossary of terms - It is extremely valuable, both for comprehension and for precision of terms utilized, to assemble the glossary. For each one of the terms, the glossary contains a short portrayal, conceivable equivalent words and reference to different terms that are contained in the actual glossary with which there is legitimate connection. A short glossary for our application appears in Figure 98. Once the different ambiguities and errors have been distinguished, they can be dispensed with by substituting more suitable terms for the erroneous ones. Where question remains, the client who gave the data must be met again or the specific record must be retrieved.

Term	Description	Synonym	Links
Trainee	Participant in a course. Can be an employee or self-employed.	Participant	Course, Company
Instructor	Course tutor. Can be freelance.	Tutor	Course
Course	Course offered. Can have various editions.	Seminar	Instructor, Trainee
Company	Company by which participant is employed or has been employed.		Trainee

Figure 98 -Glossary of Terms Example.

(Source: Atzeni, P. D. Relational Database Theory.)

Give us a chance to take a gander at the fundamental alterations to be made to our content. To start with, all the cases obviously participant must be replaced by trainee. At that point, as we have stated, place of birth (lines-4 and 17) ought to be substituted by town of birth. Additionally, we need to make unequivocal that address and phone numbers (line-5) refer to data of the trainees. The terms period employed (line-6) and assessment (line-7), must be replaced by begin date and end date, and marks out of ten, respectively. It should likewise be evident that a learner does not go to workshops (line-8) rather editions of the courses. Likewise, the term place (line-9) must be replaced with classroom, title (line-15) with professional title, and tutor (line-18) with instructor. Lastly, the title (line-

10) is a homonym of professional title of a trainee. For this situation, we could replace title with position held.

Now, we can modify our details utilizing the proposed adjustments. It is extremely valuable, in this stage, to separate the content into gatherings of homogenous expressions, alluding to a similar idea. Along these lines, we can start to force a structure on the requirements as appeared in Figure 99.

Obviously, the particular of the information itself, we additionally need to indicate that the operations to be delivered on this information. For this, we should utilize the same phrasing as that decided for the information and we should discover how frequently the different operations should be done. As we might see, this information will be imperative in the legitimate plan stage. For our own applications, the operations on the information could be as per the following:

- operation-1: embed another trainee including all his or her information (to be completed roughly 40 times each day);
- operation-2: allot a trainee to a release of a course (50 times each day);
- operation-3: embed another instructor, including all his/her information and the courses he/she is fit to instruct (twice per day);
- operation-4: leave out a qualified instructor to editions of the courses
- operation-5: show all the info on the past editions of the courses with name, start date and end date, number of instructors (10 times per day);
- operation-6: show all the course offered, with info on the instructor who are qualified to teach them (20 times per day);
- operation-7: for each instructor, find the trainees for all the courses he/ she is teaching or has already taught (5 times per week);
- operation-8: leave out a statistical analysis of all the trainees with all the info about them, about the editions of the courses they have attended and the marks they have obtained (10 times per month)

Phrases of a general nature
We wish to create a database for a company that runs training courses. We wish to hold the data for the trainees and the instructors.
Phrases relating to the trainees
For each trainee (about 5000), identified by a code, we will hold the social security number, surname, age, sex, town of birth, current employer, previous employers (along with the start date and the end date of the period employed), the editions of the courses the trainee is attending at present and those he or she has attended in the past, with the final marks out of ten.
Phrases relating to the employers of the trainees
For each employer of a trainee we will hold the name, address and telephone number.
Phrases relating to the courses
For each course (about 2oo), we will hold the name and code. Each time a particular course is given, we will call it an 'edition' of the course. For each edition, we will hold the start date, the end date, and the number of participants. For the editions currently in progress, we will hold the dates, the classrooms and the times in which the classes are held.
Phrases relating to specific types of trainee
For a trainee who is a self-employed professional, we will hold the area of expertise and, if appropriate, the professional title. For a trainee who is an employee, we will hold the level and position held.
Phrases relating to the instructors
For each instructor (about 300), we will hold surname, age, town of birth, all telephone numbers, the edition of courses taught, those taught in the past and the courses the instructor is qualified to teach. The instructors can be permanently employed by the training company or can be freelance.

Figure 99 - Structuring of Requirements.

(Source: Atzeni, P. D. Relational Database Theory.)

GENERAL CRITERIA FOR INFORMATION REPRESENTATION

As a prologue to plan strategies, we will build up some broad criteria for the interpretation of casual determinations into Entity-Relationship diagram. It must be focused on that a similar data that can be displayed in various ways. This is on the grounds that two planners can have extraordinary impression of a similar circumstance. We can however build up some broad rules for the E-R model. It is prescribed, generally speaking, that the 'applied rules' of the model ought to be taken after.

- If an idea has critical properties as well as depicts classes of articles with a self-sufficient presence, it is proper to speak to it by an entity. For illustration, alluding to the particulars for the preparation organization above, it bodes well to speak to the idea of instructor with an entity, in light of the fact that it has different attributes (surnames, age, and birthplace) and its presence is free of alternate ideas.
- If an idea has a basic structure, and has no pertinent properties related with it, it is helpful to speak to it by a credit of another idea to which it alludes. For instance, the idea of age can absolutely be spoken to as an attribute of the trainee. When all is said in done, the idea of town could have noteworthy properties. In our application, it is better to demonstrate it as an attribute, in light of the fact that, aside from the 'name', no one of its other properties is important to us.
- If the requirements contain an idea that gives a coherent connection between two (or even more) entities, this idea can be spoken to by a relationship. For case, in our application, the idea of going to a course can surely be spoken to by a relationship between the entity that speak to the trainees and the editions of courses. It must be highlighted that this is substantial just in the circumstance in which the idea does not, itself, have the attributes of the entity. An example is the idea of exam identifying the trainees and courses. This idea can be spoken to by a relationship among trainees and course if the part of intrigue is, say, just the singular student marks. Assume nonetheless, that we are likewise keen on the date, the area and the looking at board, and, most importantly, that we wish to speak to the way that a trainee who can take an exam more than one time for a similar course. For this situation the exam must be spoken to by an element connected by one-to-many relationship to the entity referring to trainees and courses.
- If at least one of the ideas is specific instances of another idea, it is helpful to speak to them by methods for a speculation. In our application, clearly the ideas of expert and representative

constitute specific cases of the idea of student and it is in this manner a smart thought to characterize a speculation between the elements that speak to these ideas.

- These criteria are substantial all in all, and are subsequently free of the picked outline procedure. For each system, it is important to make an interpretation of a detail into an E-R diagram.

CHAPTER 7

CONCEPTUAL SCHEMAS AND STRATEGIES OF DESIGN

STRATEGIES OF DESIGN

The advancement of an applied pattern in view of its detail must be considered to all goals and purposes a building procedure, and, all things considered, outline methodologies utilized as a part of different controls can be connected to it. Let us look at these systems with particular reference to the displaying of a database.

Top-Down Strategy

In this strategy, the theoretical diagram is delivered by methods for a progression of progressive refinements, beginning from an underlying composition that depicts all the requirements by methods for a couple of exceptionally dynamic ideas. The mapping is at that point step by step extended by utilizing fitting alterations that expand the detail of the different ideas. This strategy is appeared in Figure 100 where the different levels of refinement are also shown. Each one of these levels contains a diagram that depicts a similar data with an alternate level of detail. Hence, in an immaculate best down procedure, every one of the ideas introduce in the last model are available, on a basic level, at each level of refinement. Moving from the one level to the next, the outline is altered utilizing some essential changes called top-down transformation primitives. Significant cases of top-down transformation primitives are appeared in Figure 101. As we can see, these primitives work on a solitary idea of the diagram and change it into a more unpredictable structure, equipped for portraying the first idea in more detail (Teorey T. J., 1990).

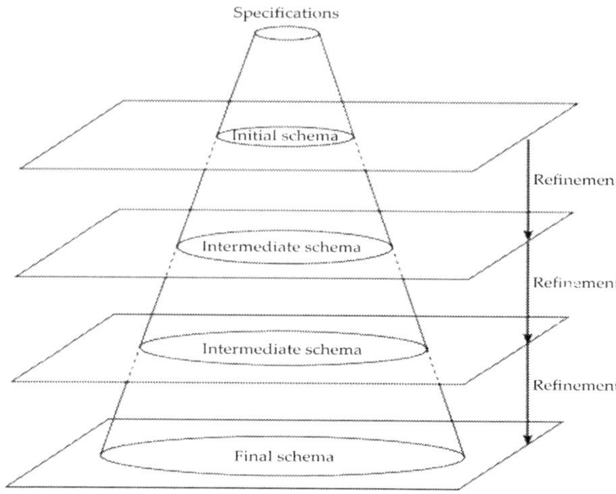

Figure 100 - Strategy Top-Down.

(Source: Senn, J. A. Analysis & Design of Information Systems.)

Transformation	Initial concept	Result
T_1 From one entity to two entities and a relationship between them		
T_2 From one entity to a generalization		
T_3 From one relationship to multiple relationships		
T_4 From one relationship to an entity withe relationships		
T_5 Adding attributes to an entity		
T_5 Adding attributes to an relationships		

Figure 101 - Transformation Primitives.

(Source: Senn, J. A. Analysis & Design of Information Systems.)

Transformation T_1. This is connected when an element depicts two extraordinary ideas coherently connected to each other. For instance, in the preparation organization application, we could have started with an element COURSE. At that point we understand this is excessively theoretical and that COURSE-TYPE (having one code and one title) ought to be particular from COURSEEDITION (having a begin date and an end date) and that these entities are connected by a relationship that we actually call TYPE.

Transformation T_2. This is connected when an entity is comprised of particular subentities. In our application, this happens when we understand that among the trainees there are the particular cases EMPLOYEE and PROFESSIONAL.

Transformation T_3. This is connected when a relationship depicts two or more extraordinary ideas connecting similar entities. For instance, in the TEACHING relationship among instructor and trainees, CURRENTTEACHING ought to be recognized from PASTTEACHING.

Transformation T_4. This is connected when a relationship depicts an idea having a self-ruling presence. For instance, if a relationship CONTRACT between CONSULTANT entity and COMPANY entity has many traits, it is better spoken to by an element connected to the others by methods for parallel connections.

Transformation T_5. This is connected for the expansion of attributes (qualities) to entities. This happens, for instance, when we refine the TRAINEE entity by including its SocialSecurityNumber, Surnames, Ages, Sex, and TownOfBirth attributes.

Transformation T6. This is connected to add attributes to relationships in a comparable approach to Transformation T_5.

The benefit of the top-down strategy is that the planner can begin with a full portrayal of the requirements, despite the fact that a few points of interest are missing. Clearly, this is conceivable just on the off chance that we have a worldwide perspective of all the pertinent ideas. This is greatly troublesome when managing complex cases.

Bottom-Up Strategy

In this strategy, the underlying particulars are decomposed into littler and littler segments, until the point that every part portrays a rudimentary piece of the particulars. Now, the different segments are spoken to by basic applied mappings that can likewise comprise of single ideas. The different outlines in this way acquired are then amalgamated until a last applied composition is come to. This strategy appears in Figure 102, which demonstrates the decomposition of the requirements, the consequent development of the essential E-R diagram, and the last stage in which the basic compositions are coordinated. As opposed to the top-down strategy, the different ideas exhibit in the last composition is logically presented. For this situation, the last mapping is acquired by methods for some rudimentary changes, in here the bottom-up transformation primitives. In Figure 103, the fundamental bottom-up transformation primitives are appeared. As we can see, these primitives bring into the pattern new ideas that were not present beforehand and are equipped for portraying parts of the application that have not been considered some time recently.

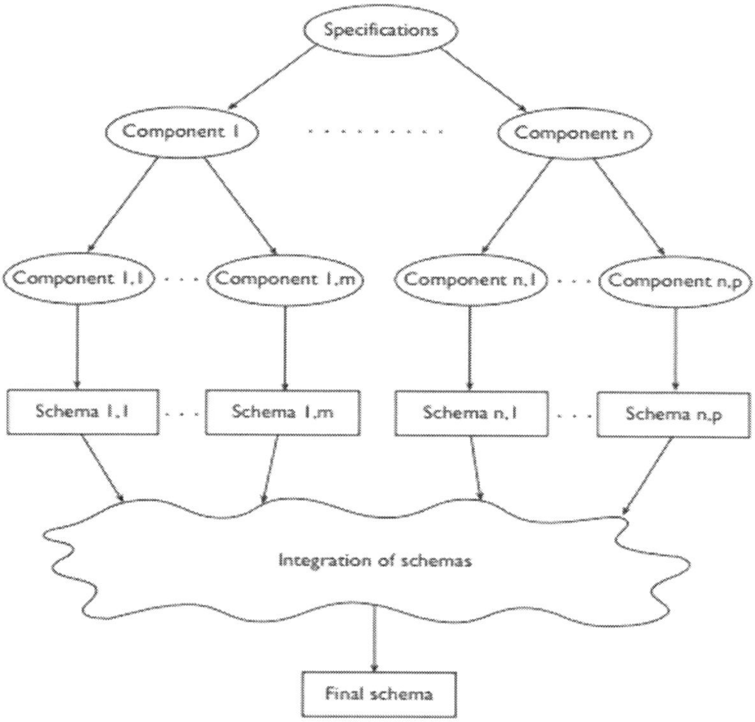

Figure 102 – Bottom-Up Strategy.

(Source: Senn, J. A. Analysis & Design of Information Systems.)

Transformation T_1. This is connected when a class of items with attributes in like manner is distinguished in the determinations. For instance, in the application for the preparation organization, this can happen when we distinguish the COURSECLASS entity (held in a specific classroom at a specific time) from the course detail.

Transformation T_2. This is connected when an intelligent relationship between two elements is recognized in the determinations. In our applications, this could happen when we recognize the QUALIFICATION relationship between the INSTRUCTOR entity and COURSE entity.

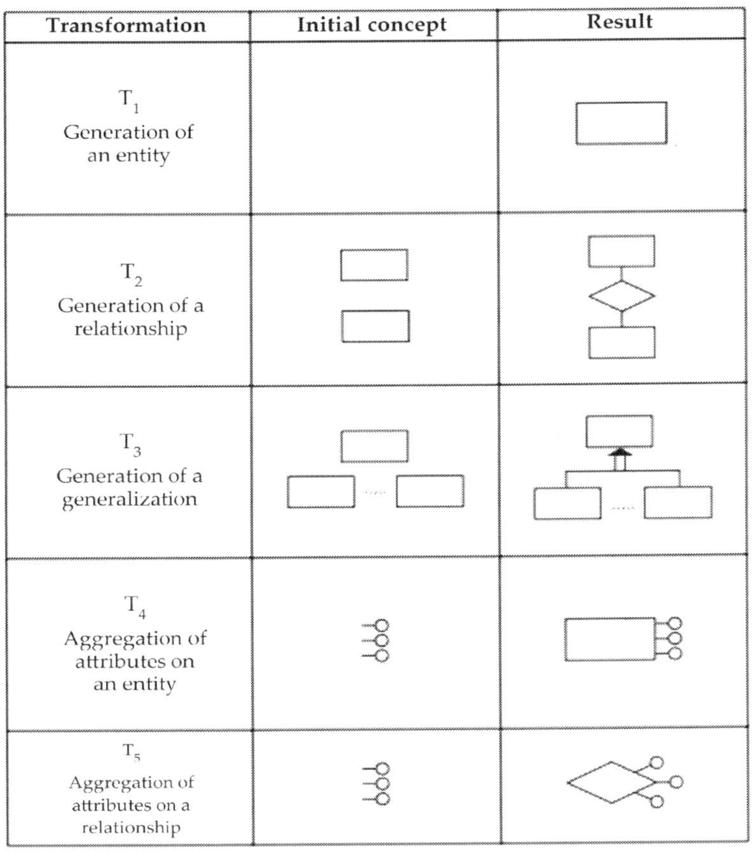

Figure 103 – Bottom-Up Primitives.

(Source: Senn, J. A. Analysis & Design of Information Systems.)

Transformation T_3. This is connected when a speculation between substances is recognized in the determination. For instance, in our type of application this thing can happen when we comprehend that the INSTRUCTOR entity is a speculation of the PERMANENT entity and FREELANCE entity.

Transformation T_4. This is connected when we distinguish an entity that can be viewed as a total of an aggregation of attributes. For instance, this happens in our application in the event that we distinguish the TRAINEE entity from the SocialSecurityNumber, Surnames, Ages, Sex, and TownOfBirth attributes.

Transformation T_5. This is connected in a path like TransformationT_4, when a relationship is recognized that can be viewed as an aggregation of attributes.

The upside of the bottom-up strategy is that it permits the deterioration of an issue into basic parts, which are frequently effortlessly distinguished, and whose outline can be taken care of by various creators if essential. It subsequently fits work attempted in gatherings or subdivided inside a gathering. Then again, its disservice is the way that it requires the incorporation of numerous applied patterns, an operation that frequently exhibits troubles (Ceri S. , 1984).

Inside-Out Strategy

This system can be viewed as a specific sort of bottom-up strategy. It starts with the recognizable proof of just a couple of critical ideas and, based on these, the outline continues, spreading outward 'radially'. That is, first the ideas closest to the underlying ideas are spoken to, and we at that point move towards those further away by methods for "route" through the determination. A case of back to front advancement of a calculated mapping is appeared in Figure 108 with reference to an illustration found in the past section. In this chart, the different ranges show a conceivable sequential advancement of the mapping. Note that the EMPLOYEE entity was recognized firstly, alongside its attributes.

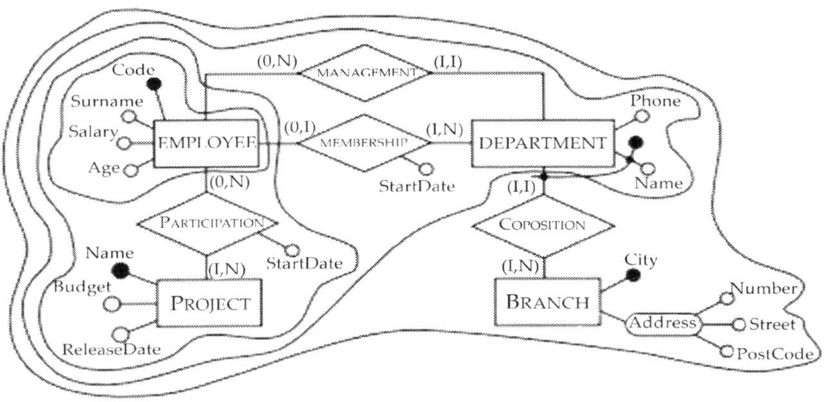

Figure 104 – Inside-Out Strategy.

(Source: Senn, J. A. Analysis & Design of Information Systems.)

On this premise, the employee participation in ventures was then spoken to, together with the properties of the tasks. Following this, the connection between the representatives and the organization divisions was dissected, recognizing the relationships MANAGEMENT and MEMBERSHIP and the DEPARTMENT entity. At long last, in view of this last element, the branches of the organization were spoken to (BRANCH substance with its attributes), together with the reality that the different offices have a place with the branches (COMPOSITION relationship).

This technique has the upside of not requiring incorporation steps. On the other hand, every now and then, it is important to look at all of the determination searching for ideas not yet spoke to and to portray these new ideas in detail.

Mixed Strategy

As is regularly the case, each of the strategies has both positive what's more, negative viewpoints. Along these lines, we can devise a blended system that tries to join the benefits of top-down strategy with those of bottom-up and inside-out. The originator disintegrates the necessities into various segments, as in the base up procedure, however not to the degree where all the ideas are isolated. In the meantime he or she characterizes a skeleton construction containing, at the dynamic level, the central ideas of the application.

This skeleton pattern gives a brought together, albeit manufactured, perspective of the entirety outline and supports the incorporation of the patterns grew independently. For instance in Figure 107, we demonstrate a conceivable model in preparing organization application. From a straightforward investigation of the prerequisites in Figure 108, we can quickly distinguish three main ideas that can be spoken to by elements: the trainees, the courses, and the instructors. There are connections between these entities that we can expect to be depictions of the participation of the trainees to the editions of the courses and of the showing exercises embraced by the teachers of the actual courses. At this point we can proceed onward, inspecting independently these primary ideas and can continue with progressive refinements (along these lines following the top-down strategy) or expanding the different parts of the model with ideas that are not yet spoken to (along these lines following the base up procedure).

Figure 105 - Training Company Schema.

(Source: Senn, J. A. Analysis & Design of Information Systems.)

The mix system is likely the most adaptable of those we have seen, since it loans it well to different requirements: that of subdividing a perplexing issue into littler ones and that of continuing by progressive refinement. Truth be told, this procedure additionally envelops the inside-out strategy, which is just a variety of the bottom-up procedure. It is entirely characteristic, amid the base up advancement of a sub-component of the venture, to continue from the inside-out keeping in mind the end goal to speak to the parts of the determination of the database that are not shown yet. It must be said that, in all the practical examples, the mixed strategy is the only that can actually be adopted. It is often necessary to start the design before all the info requirements are acquirable.

QUALITY OF A CONCEPTUAL DESIGN

A "quality" conceptual design is ought to have various properties, which can be utilized to check the nature of the model itself. Give us a chance to break down the most essential of these properties and perceive how we can check them amid the conceptual design plan.

Accuracy

A conceptual design is accurate when it utilizes appropriately the builds made accessible by the reasonable model. Likewise with programming languages, the blunders can be syntactic or semantic. The first identifies with unlawful utilization of a design, like in determining a speculation between relationships as opposed to between entities. The second identifies with the utilization of a design that does not take after its definition. For examples is the utilization of a relationship to depict the way that an entity is a specialization of another.

he accuracy of a design can be confirmed by assessment, contrasting the ideas show in the mapping with the requirements and with the meanings of the builds of the theoretical model utilized.

Culmination

A conceptual design is finished when it incorporates ideas that speak to every one of the information requirements and take into consideration the execution of all the operations incorporated into the operational requirements. The culmination of a design can be confirmed by seeing that every one of the requirements on the information are defined by some idea show in the construction, and that every one of the ideas engaged with an operation that come by "exploring" over the design.

Lucidness

A conceptual design is meaningful when it speaks to the requirements in a way that is common and straightforward. Thus, the model must act naturally logical, for instance, by picking reasonable names for ideas. Additionally, clarity relies upon absolutely stylish criteria: the appreciation of a construction is made less demanding on the off chance that we draw it with steady measurements for its parts. A few recommendations for making the design more comprehensible are as follows:

- mastermind the design on an actual grid, picking as focal components those that have most (connections) with others;
- utilize just even and vertical lines and attempt to keep crossing points to a least; and
- orchestrate parent entities substances over the particular offspring entities;

The comprehensibility of an outline can be checked via doing appreciation tests with the clients.

Insignificance

An outline is negligible when every one of the determinations on the information is spoken to just once in the mapping. A mapping is in this way not negligible at the point when there are redundancies, ideas that can be gotten from others. An average wellspring of repetition in an E-R outline is the nearness of cycles caused by the nearness of connections as well as speculations.

A repetition is not generally undesirable, but can be the consequence of exact plan choices. Regardless, these circumstances require dependably to be reported. The insignificance of a pattern can be confirmed by examination, checking whether there exist ideas that can be erased from the outline without bargaining its culmination.

In the following area we will perceive how the confirmation of the nature of a conceptual design can be consolidated into a far reaching strategy.

COMPREHENSIVE METHODS FOR CONCEPTUAL DESIGN

In this area, we will aggregate up all that we might have said regarding theoretical database design. With respect to the outline systems we have seen, we should stretch that by and by it is uncommon for a venture to continue dependably in a top-down or in a bottom-up strategy. Freely of the picked procedure, in reality there is dependably a need to alter the diagram being built utilizing both top-down changes that refine ideas officially present and bottom-up changes that include ideas not officially present. We will along these lines demonstrate a strategy for conceptual design with reference to the mix system, which, as we have stated, incorporates the others as exceptional cases. The method is comprised of the accompanying strides.

- Investigation of requirements
    - Development of a glossary of actual terms.
    - Analyse the requirements and wipe out any ambiguities.
    - Arrange the requirements by groups.
- Fundamental stride
    - Identify the most pertinent ideas and speak to them in a skeleton outline.
- Deterioration venture (to be utilized if proper or essential).
    - Decompose the requirements with reference to the ideas introduce in the skeleton pattern.
- Iterative stride: to be rehashed for every one of the patterns until each detail is spoken to.
    - Refine the ideas in the composition, in view of the requirements.
    - Add new ideas to the composition to depict any parts of the requirements not yet spoken to.
- Incorporation venture (to be completed if step 3 has been utilized).
    - Integrate the different sub-schemas into a general outline with reference to the skeleton outline.
- Quality investigation
    - Verify the accuracy of the pattern and complete any important rebuilding.
    - Verify the culmination of the mapping and do any vital rebuilding.
    - Verify the insignificance, list the redundancies and if essential rebuild the construction.
    - Verify the coherence of the composition and do any fundamental rebuilding.

Note that if Step 3 and Step 5 are discarded, and if in Step 4 just refinements are made, we have an immaculate top-down strategy. Then again, if the essential stride is not done and if in the iterative stride new ideas are included, we are utilizing an immaculate bottom-up strategy. In the bottom-up changes, we can continue as indicated by the inside-out system.

In the technique appeared, we have just quickly said an imperative movement that goes with each plan stage: the documentation of the patterns.

At last, a couple of remarks on the investigation of the nature of the plan are required. To start with, it must be expressed that quality investigation ought not be consigned to the finish of the theoretical outline: in fact this is a parallel action, which ought to be done routinely amid the improvement of a theoretical composition. Moreover, it is extremely fragile, since it frequently requires rebuilding to be completed keeping in mind the end goal to cure "blunders" made in past stages. It is important to give careful consideration to ideas of the blueprint that have specific properties: for instance, entities without attributes, grouping of ideas that frame cycles, over-complex speculation chains of command or especially convoluted segments of the construction.

A CASE OF CONCEPTUAL DESIGN

Let us see an entire case of reasonable plan, again with reference to our preparation organization. We have just completed the primary stage of the strategy depicted above and we have demonstrated a conceivable skeleton mapping in Figure 106. With reference to this construction, we can, choose to break down independently the determinations that identify with the trainee, courses and the instructor, and to continue inside-out for each one of them.

We will now complete the iterative stride, managing first with the trainees. Of these, two sorts are instantly identifiable: the professionals and the employees. These entities can be spoken to as a specialization of the TRAINEE substance: the subsequent speculation is add up to. Now, the businesses of the trainees should be spoken to. This should be possible by presenting the EMPLOYER entity, which is connected by a relationship to the EMPLOYEE entity. In the event that we dissect the requirements, we see that we have to speak to two particular ideas: over a significant time span work. We will choose accordingly, to subdivide this relationship into two relationships: PAST EMPLOYMENT and PRESENT EMPLOYMENT. The first has a StartDate and an EndDate and is related to the TRAINEE entity, while the second has just a StartDate and is related to the EMPLOYEE entity. Watch that the TRAINEE entity has two identifiers, the code and SSN. Note likewise that the ProfessionalTitle attribute is discretionary, in that the cardinality expresses that this data can be zero.

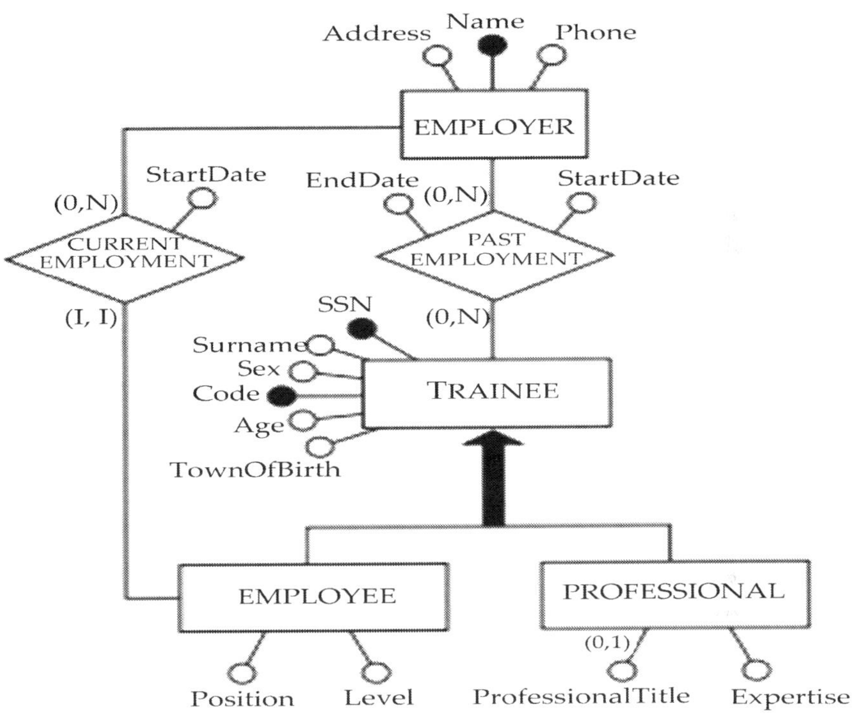

Figure 106 - The Refinement of One Portion.

(Source: O'Neil, P. Database: Principles, Programming, Performance.)

For the instructor, we have to recognize those who are PERMANENT or FREELANCE. This should be possible normally with an aggregate speculation of which INSTRUCTOR is the parent entity. We would then be able to include the characteristics Surname, Age, TownOfBirth, and Phone to the INSTRUCTOR entity. The last trait is multi-valued since an instructor can have more than one phone number and we wish to speak to them all. Note that the accessible characteristics given no characteristic identifier for the INSTRUCTOR. Here, we can choose to utilize the SSN of the instructor, regardless of the possibility that this data is not in the requirements. As an option, we could have presented an interior code, utilized just for this reason.

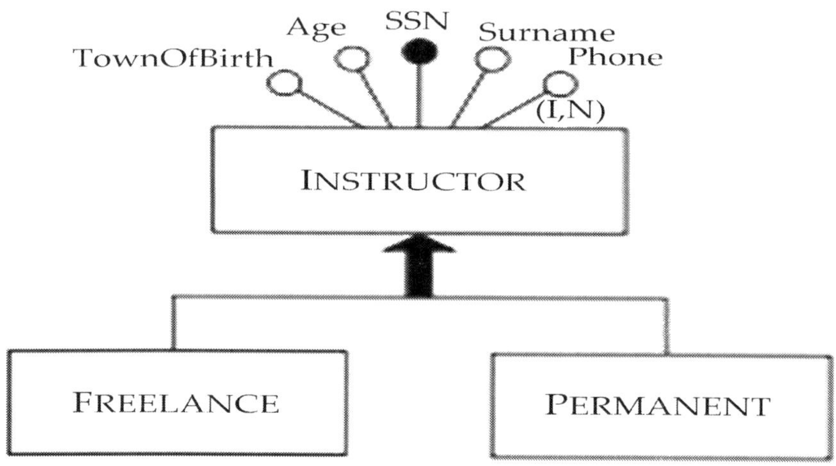

Figure 107 - The Refinement of One Other Portion.

(Source: O'Neil, P. Database: Principles, Programming, Performance.)

CASE TOOLS FOR DATABASE PLAN

Database configuration is an unpredictable action that is frequently troublesome or difficult to complete physically. The procedure can be made less demanding by utilizing universally useful altering programs with graphical interfaces for making tables and charts.

There are, programming bundles explicitly intended for the creation and improvement of databases. These frameworks have a place with the class of case (Computer-Aided Software Engineering) devices and offer help for the fundamental periods of the improvement of a database (conceptual, logical, and physical design).

The functionalities shift generally starting with one bundle then onto the next, however there are some fundamental components that are available in a pretty much broad frame on the whole systems:

- a graphical-interface with which it is conceivable to control E-R models diagrammatically;
- an information word reference, which stores data on the different parts of the diagram (elements, characteristics, connections, trustworthiness requirements, and so forth.); and
- a progression of coordinated apparatuses, which complete, either consequently or through connection with the client, particular plan assignments (programmed design of graphs, confirmation of accuracy and culmination, quality examination of a mapping, programmed generation of DDL code for the production of a database, and so forth.).

Numerous systems are coordinated specifically with database administration frameworks. Different systems additionally offer help for necessities investigation. Still others give libraries of predefined non-specific tasks that can be utilized as a beginning point for another venture. With particular respect to calculated outline, it is for the most part conceivable to take after the techniques proposed in the areas above, notwithstanding when these frameworks are utilized. A large number of them make it conceivable to utilize a best down procedure, permitting the fractional detail of ideas of the pattern and the continuous refinement of them. For instance, we can characterize an element without determining characteristics or identifiers. There are as yet different systems that enable perspectives over a pattern to be characterized and controlled independently, consequently proliferating to the mapping adjustments made on the view, accordingly continuing in a bottom-up way.

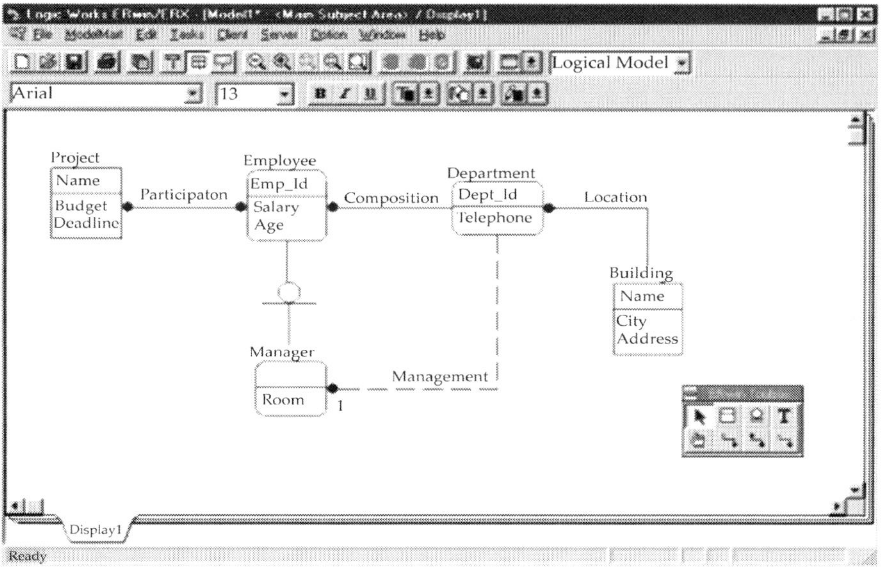

Figure 108 - CASE Tool.

(Source: O'Neil, P. Database: Principles, Programming, Performance.)

Figure 108 demonstrates a conceptual design oversaw by a standout among the most prevalent tools. Note that the documentation used to depict the builds of the E-R model is not the same as that utilized as a part of this section. Specifically the properties are specifically spoken to inside the entity, isolating the identifiers (primary key) from the other attributes. The lines speak to relationships and specific images on the lines are utilized to express cardinality requirements. Speculations are spoken to by lines isolated by a unique image (connection between EMPLOYEE and MANAGER). This portrayal does not permit the determination of characteristics of relationships (Batini, Conceptual Database Design, an Entity-Relationship Approach, 1992).

This case demonstrates an outstanding issue that must be stood up to when utilizing a CASE tool for database configuration: there are institutionalizations neither of the documentations utilized, nor of the reasonable model, and every system considerably receives a particular rendition of the E -R model. There are systems that oversee speculations of just a specific sort, others that do not oversee them by any means, still others that oversee just paired connections. Moreover, every one of the items utilizes unique documentations to speak to similar develops. In this way, extensive exertion is frequently required with respect to the originator to adjust his or her learning of models and strategies to the attributes of the picked item.

CHAPTER 8

LOGICAL DESIGN

The point of logical design is to build a coherent construction that effectively and effectively speaks to the majority of the data portrayed by an Entity-Relationship model delivered amid the applied plan stage. This is not only a straightforward interpretation starting with one model then onto the next for two reasons. Initially, there is no nearby correspondence between the models included on the grounds that not every one of the builds of the Entity-Relationship model can be deciphered normally into the relational model. For instance, while an element can without much of a stretch be spoken to by a connection, there are different alternatives for the speculations. Also, the point of reasonable outline is to speak to the information precisely from an abnormal state, computer independent perspective. Consistent outline is rather the reason for the genuine usage of the application, and must consider, similarly as conceivable, the execution of the last item. The pattern should in this manner be rebuilt so as to make it the execution of the anticipated operations as proficient as could be allowed. In aggregate, we should design an assignment that is not just an interpretation (from the applied model to the logical one) yet additionally a rearrangement. Since the revamping can generally be managed autonomously of the coherent model, it is typically useful to separate the sensible plan into two stages, as appeared in Figure 109.

- Restructuring of the Entity-Relationship schema - which is free from the picked consistent model, and depends on the criteria for the improvement of the pattern and the rearrangements of the accompanying stride.

- Translation into the coherent schema - which alludes to a particular logical show and can incorporate a further streamlining, in view of the components of the sensible model itself. The contribution for the initial step is the conceptual schema created in the going before stage and the assessed database stack, regarding the measure of information and the operational requirements. The outcome acquired is a reconstructed E-R schema, which is never again a conceptual schema in the strict feeling of the term, in that it does constitute a portrayal of the information that considers usage viewpoints.

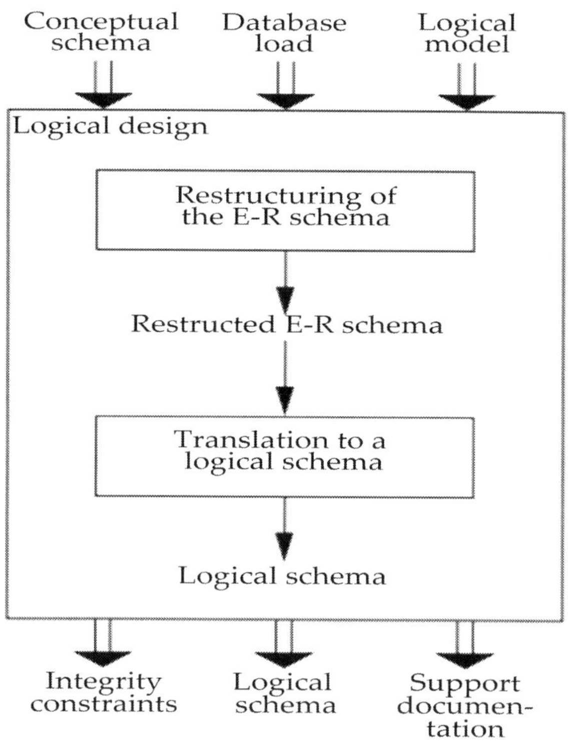

Figure 109 - Design of logical Database.

(Source: O'Neil, P. Database: Principles, Programming, Performance.

This diagram and the picked legitimate model constitute the contribution of the second step, which creates the coherent pattern of our database. In this progression, we can likewise do quality controls on the mapping and, conceivably, further enhancements in view of the attributes of the coherent model. An examination method utilized for the relational model (called standardization) will be displayed independently in the following part. The last legitimate mapping, the respectability imperatives characterized on it and the pertinent documentation, constitute the last result of sensible outline.

In the rest of this part, we will display the two stages that make up the sensible outline of a database. We will initially examine the systems that can be utilized to break down the productivity of a database by alluding to its logical schema.

E-R SCHEMAS PERFORMANCE ANALYSIS

An E-R schema can be changed to streamline some execution markers. We utilize the term pointer, on the grounds that the effectiveness of a database cannot be absolutely assessed with reference to a reasonable mapping. The reason is that the genuine conduct is additionally subject to physical perspectives that are most certainly not germane to a calculated portrayal. It is conceivable, to do assessments of the two parameters that impact the execution of any programming. These are:

- cost of actual an operation: this is assessed as far as the quantity of instances of entities and relationships that are gone by to actually execute an operation on the database; this is a coarse measure and it will now and again be important to allude to more definite criteria; and
- capacity necessity: this is assessed as far as number of bytes, and is important to store the information depicted by the composition.

Keeping in mind the end goal to think about these parameters, we require the following data.

- Volume of information. That is:
    - number of events of every element and relationship of the diagram;
    - size of each characteristic.
- Operation qualities. That is:
    - kind of operation (intelligent or group);

- recurrence (normal number of executions in a specific time traverse);
- information included (substances and additionally connections).

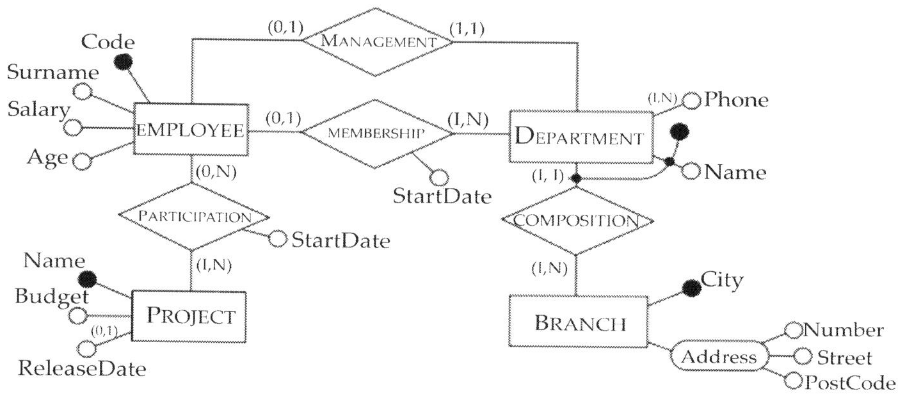

Figure 110 - E-R Schema.

(Source: Abiteboul, S. H. Foundations of Databases.)

Typical operations for this pattern can be:

- operation-1: dole out an employee to a project;
- operation-2: discover the information for an employee, for the department in which he or she works and for the projects in which he or she participates;
- operation-3: discover the information for every one of the employees of a specific department;
- operation-4: for each branch, discover its specializations with the surnames of the head and the rundown of the employees in every department.

Despite the fact that the operations above might look distorting as for the real database stack, we can take note of that database operations take after the so called 'eighty-twenty full rule'. This rule states that 80% of the heap is created by 20% of the operations. This reality enables us to focus just on a few operations and still give a sufficient sign of the workloads for the resulting examination.

The volume of information and the general attributes of the operations can be portrayed by utilizing tables, for example, those in Figure 111. In the table of volumes, every one of the ideas of the diagrams appears (elements and connections) with their evaluated volumes. In the actual table of operations appeared, for each operation, the normal recurrence and an image that demonstrates whether the operation is intelligent (I) or group (B). In the volumes table, the number of events of a relationship relies upon two parameters. These are (i) the volume of the substances associated with the relationship and (ii) the number of times an event of these substances partakes by and large in an event of the relationship. The last depends thusly on the cardinalities of the relationship. For instance, the quantity of events of the Organization relationship is equivalent to the quantity of divisions, since the cardinalities manage that every division has a place with one and just a single branch. Then again, the quantity of events of the relationship Participation is minimal, not as much as the quantity of Employee, since a couple of employees have a place with no Department. Lastly, if an employee is included in three projects, we have 2000×3=6000 events for the PARTICIPATION relationship (and along these lines 6000 ÷ 500 = 12 employees considered for each extend) (Teorey T. J., 1982).

Table of volumes		
Concept	**Type**	**Volume**
Branch	E	10
Department	E	80
Employee	E	2000
Project	E	500
Composition	R	80
Membership	R	1900
Management	R	80
Participation	R	6000

Table of operations		
Operation	**Type**	**Frequency**
Op 1	I	50 per day
Op 2	I	100 per day
Op 3	I	10 per day
Op 4	2	500 per day

Figure 111 - Volume Table and Also Operations Table.

(Source: Abiteboul, S. H. Foundations of Databases.)

For every operation we can, in addition, depict the information required by implies of a route pattern that comprises of the piece of the E-R model applicable to the operation. On this diagram, it is helpful to draw the 'logical way' that must be taken after to get to the required data. A case of a route pattern is proposed in Figure 112 with reference to operation 2. To get the required data, we start with the EMPLOYEE entity and we access his area of expertise by methods for the MEMBERSHIP relationship, also, to the tasks in which he is included by methods for the PARTICIPATION relationship.

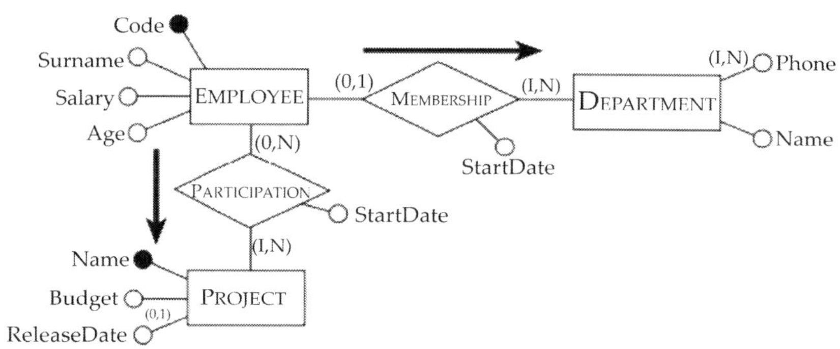

Figure 112 - Navigation Schema.

(Source: Abiteboul, S. H. Foundations of Databases.)

When this data is accessible, we can appraise the real cost of an operation into the database by checking the quantity of events of the entities and relationships important to do the operation. Look at operation-2 once again. As indicated by the route pattern, we should first get to an event of the EMPLOYEE entity all together to get to an event of the PARTICIPATION relationship and, by this implies, to an event of the DEPARTMENT entity. Following this, to acquire the information of the PROJECT on which he or she is working, we should access by and large three events of the PARTICIPATION relationship (since we have said that by and large a representative chips away at 3 projects). Through this, we access three events of the PROJECT entity. This can be shown in a table of accesses shown in Figure 113.

Table of accesses			
Concept	**Type**	**Accesses**	**Accesses**
Employee	Entity	1	R
Employment	Relation	1	R
Department	Entity	1	R
Participation	Relation	3	R
Project	Entity	3	R

Figure 113 - Access for Operation-2.

(Source: Abiteboul, S. H. Foundations of Databases.)

E-R RECONSTRUCTION

The reconstruction tasks of an E-R schema can be sub-separated into a progression of errands to be completed in succession (look at Figure 114).

- Analysis of redundancies chooses whether to erase or hold conceivable redundancies exhibit in the composition.
- Removing generalizations replaces every one of the speculations in the construction by different builds.
- Partitioning and merging of entities and relations chooses regardless of whether is it advantageous to parcel ideas in the mapping into more than one idea or to combine a few separate ideas into a solitary one.
- Selection of primary identifiers picks an identifier for those elements that have more than one.

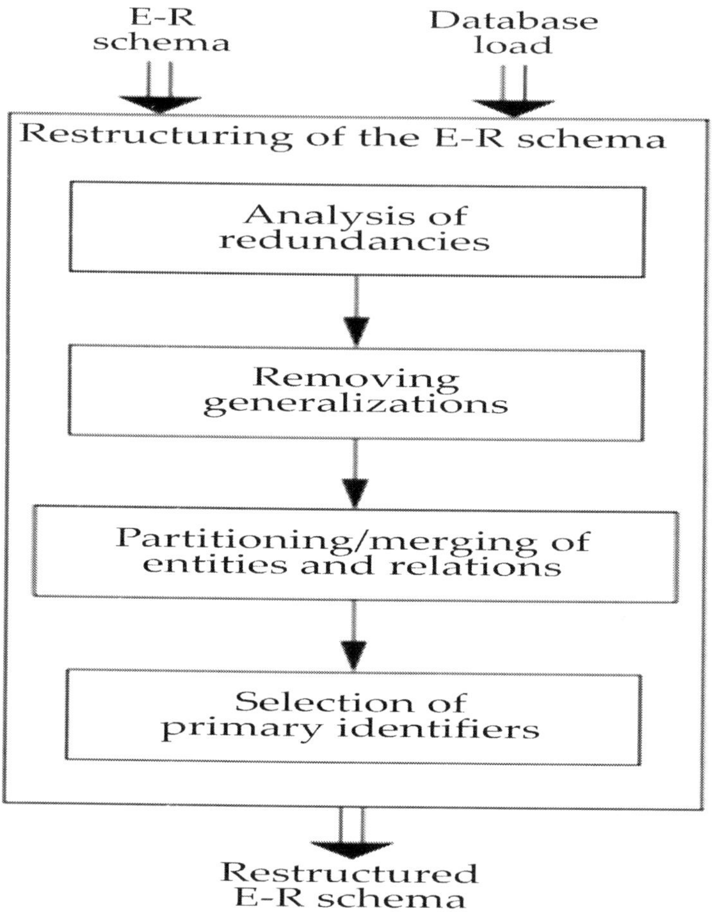

Figure 114 - Reconstructing Tasks.

(Source: Abiteboul, S. H. Foundations of Databases.)

Analysis of Redundancies

A repetition in a reasonable pattern relates to a snippet of data that can be determined (acquired by a progression of recovery operations) from other information. An Entity-Relationship schema can contain different types of redundancies. The most incessant illustrations are as per the following.

- Attributes whose esteem can be inferred, for every event of an element (or, a relationship), from estimations of different characteristics of a similar event. For instance, the main construction in Figure 115 comprises of an INVOICE entity in which one of the attributes can be derived from the others by methods for math operations.
- Attributes that can be gotten from characteristics of different entities (or relationships), for the most part by methods for total capacities. A case of such a repetition is available in the second mapping in Figure 115. In this diagram, the quality TotalAmount of the PURCHASE entity is an derived one. It can be processed from the estimations of the quality Price of the PRODUCT entity, by summing the costs of the results of which a buy is made up, as indicated by the COMPOSITION relationship.
- Attributes that can actually be gotten from operations of checking events. For instance, in the third construction in Figure 115, the NumberOfInhabitants attribute of a town can be inferred by checking the events of the RESIDENCE relationship in which the towns takes an interest. This is really a variation of the past case, which is talked about independently, as it happens often.
- Relationships that can be gotten from different connections in the presence of cycles. The last mapping in Figure 115 contains a case of this sort of repetition: the TEACHING relationship among students and lectures can be gotten from the ATTENDANCE and ASSIGNMENT relationships. It must be unmistakably expressed that the presence of the cycles does not really produce redundancies. On the off chance that, for instance, rather than the TEACHING relationship, this outline had contained a relationship SUPERVISION speaking to the connection among students and lectures, then the composition would not have been an excess (Teorey T. J., 1986).

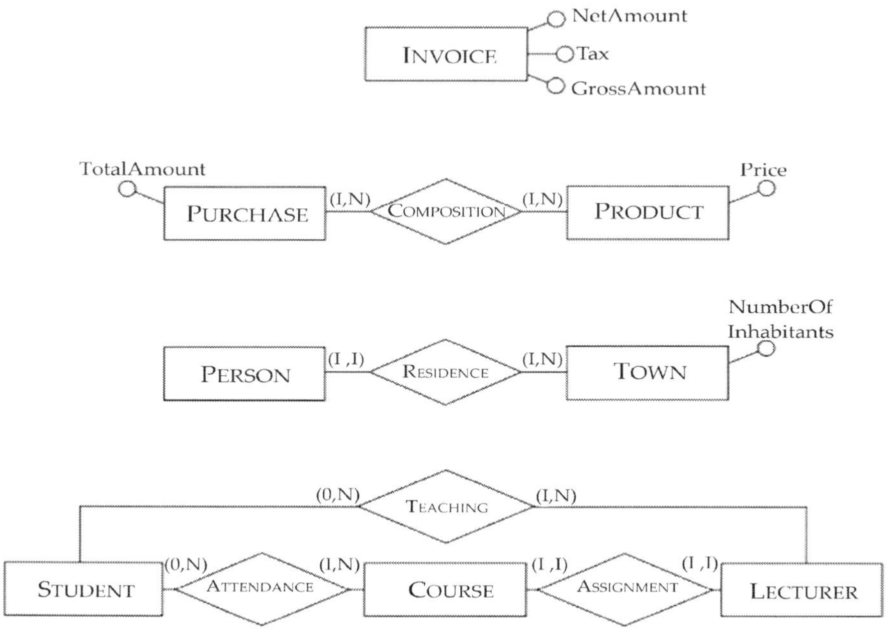

Figure 115 – Redundancies Schemas Examples.

(Source: Ramakrishnan, R. Database Management Systems.)

The presence of an inferred snippet of data in a database displays a preferred standpoint and a few drawbacks. The best position is a lessening in the number of accesses important to get the determined data. The disservices are a bigger stockpiling prerequisite (which is insignificant cost) and the need for doing extra operations in request to stay up with the latest. The choice to keep up or erase an excess is made by looking at the cost of operations that include the excess data and the capacity required, on account of essence or nonappearance of repetition.

Utilizing a viable illustration, let us take a gander at how the assessment apparatuses portrayed above can be utilized to settle on a choice of this sort. Consider the construction about individuals and towns in Figure 115 and envision that it alludes to a provincial constituent move application for which the accompanying fundamental operations are characterized:

- operation 1: include another person with the his or her town of residence.

- operation 2: print every one of the information of a town (counting the quantity of residents). For the application, the results is presented in Figure 116.

Table of volumes		
Concept	**Type**	**Volume**
Town	E	200
Person	E	1000000
Residence	R	1000000

Table of operations		
Operation	**Type**	**Frequency**
Op 1	1	50 per day
Op 2	1	2 per day

Figure 116 - Volumes and Operations for the Schema.

(Source: Ramakrishnan, R. Database Management Systems.)

Let us initially attempt to assess the pointers of execution on account of presence of repetition (trait NumberOfInhabitants in the TOWN substance).

Expect that the quantity of residents of a town requires 4 bytes. We can see that the repetitive information requires 4×200=800 bytes, which is, not exactly one K-byte of extra stockpiling. Let us evaluate the cost of the operations. As depicted in the get to tables in Figure 117, operation 1 requires the following. A compose access to the PERSON entity (to include another person), a compose access to the RESIDENCE relationship (to include another person-town match) lastly a read access (to locate the important town) and another compose access to the TOWN entity (to refresh the quantity of occupants of that event). This is altogether rehashed 500 times each day, for an aggregate of 1500 compose access and 500 read access. The cost of operation 2 is practically unimportant, as it requires a solitary read access to the TOWN entity to be rehashed 2 times a day. Assuming a compose access to cost 2 times as much as a real accesses; we have a sum of 3500 gets to a day when there is excess information.

Table of accesses in presence of redundancy			
Operation 1			
Concept	**Type**	**Acc.**	**Type**
Person	E	1	W
Residence	R	1	W
Town	E	1	R
Town	E	1	W

Operation 2			
Concept	**Type**	**Acc.**	**Type**
Town	E	1	R

Table of accesses in absence of redundancy			
Operation 1			
Concept	**Type**	**Acc.**	**Type**
Person	E	1	W
Residence	R	1	W

Operation 2			
Concept	**Type**	**Acc.**	**Type**
Town	E	1	R
Residence	R	5000	R

Figure 117 - Accesses for the Schema on the Electoral Roll of Data.

(Source: Ramakrishnan, R. Database Management Systems.)

For operation 1 we require a compose access to the PERSON entity and a compose access to the RESIDENCE relationship for the sum of 1000 compose gets to every day.

(There is no compelling reason to get to the TOWN entity since there is no inferred data). Anyway, for operation-2 we require a read access to the TOWN entity (to acquire the information for the town), which we can disregard, and 5000 read accesses into the RESIDENCE relationship (acquired by partitioning the number of persons by the quantity of towns) to figure the quantity of residents of this town. This gives an aggregate of

10,000 read accesses every single day.

Numbering double the compose accesses; we have an aggregate of 12,000 accesses to every single day at the point when there is no excess information. Along these lines, roughly 8,500 gets to additional everyday are required where there is not any repetitive information with a specific end goal to spare an unimportant K-byte. This relies upon the way that the read accesses expected to process the determined information are considerably more than the compose accesses expected to keep the inferred information forward.

Removing Generalizations

The relational model does not permit the immediate portrayal of generalizations of the E-R model. We require, in this way, to change these builds into different designs that are less demanding to interpret. The E-R model for which the interpretation into the relational model is simple, are entities and relationships.

To speak to a generalization utilizing entities and relationships, there are basically three conceivable choices.

- Crumple the offspring entity into the parent entity. The entities E1 and E2 are erased and their real properties are added to the actual parent entity E0. To this entity, a further A-type attribute is included, which serves to recognize the "sort" (E1 or even E2) of an event of E0. For instance, a speculation between the parent element PERSON and the offspring entities MAN and WOMAN can be fell into the PERSON entity by adding to it the Sex attribute. Note that the A11 and A21 attributes can accept invalid values (since they are inapplicable) for a few events of E0. Furthermore, the R2 relationship will have a base cardinality equivalent to 0 for the E0 entity (since the cvents of E2 are just a subset of the events of E0).
- Crumple the parent entity into the child entities. The parent entity E0 is erased and, for the property of legacy, its qualities, its identifier and the connections to which this entity was included, are added to both the child entities E1 and E2. The connections R11 and R12 speak to individually the limitation of the relationship R1 on the events of the substances E1 and E2.

Consider, for instance, a speculation between the substances PERSONs, having Surnames and Age as properties and SSN (Social-Security-Number) as an identifier, and the substances MAN and WOMAN. On the off chance that this is rebuilt along these lines, at that point the properties Surnames and Age and the identifier SSN are added to the two, the substances MAN and also WOMAN.

- Substitution of the speculation with connections. The speculation is changed into 2 balanced connections that connection the parent substance with the kid elements E1 and E2. There are no exchanges of qualities or relationship and the elements E1 and E2 are distinguished remotely by the substance E0. Extra requirements hold in the new pattern: every event of E0 can't take an interest in both RG1 and RG2; also, if the speculation is finished, every event of E0 must take an interest in precisely one of RG1 and RG2.

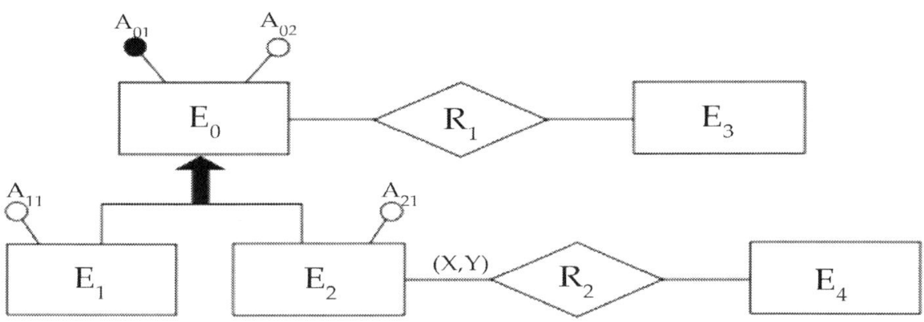

Figure 118 - Schema with Generalization.

(Source: Ramakrishnan, R. Database Management Systems.)

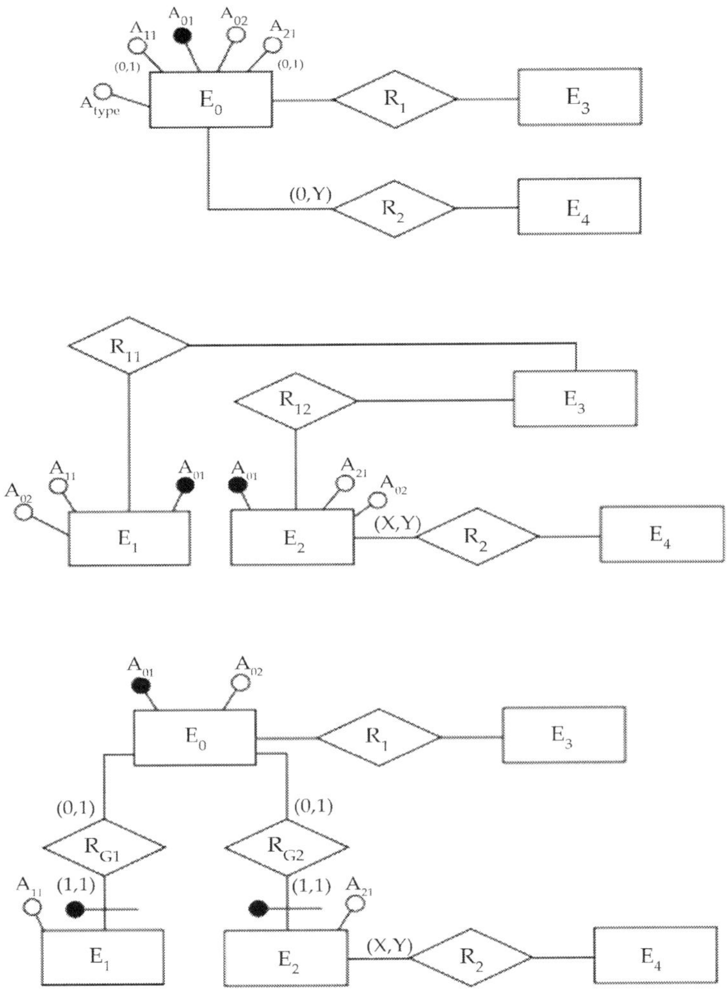

Figure 119 - Restructurings of the Schema.

(Source: Ramakrishnan, R. Database Management Systems.)

The decision among the different alternatives can be made in a way like that utilized for inferred information. That is, by considering the focal points and detriments of each of the conceivable decisions regarding capacity required and cost of the operations included. We can, be that as it may, build up some broad rules.

- Option-1: is advantageous when the operations include the events and the traits of E0, E1 and E2 pretty much similarly. For this situation, regardless of the possibility that we squander capacity for the nearness of invalid esteems, the decision guarantees less gets to contrasted with the others in which the events and the traits are disseminated among the different substances.
- Option-2: is conceivable just if the speculation is add up to, generally the events of E1 that are events of neither E1 nor E2 would not be spoken to. It is helpful when there are operations that allude just to events of E1 or of E2, thus they make refinements between these substances. For this situation, stockpiling is spared contrasted with Option-1, in light of the fact that, in rule, the properties never expect invalid esteems. Further, there is a decrease of accesses contrasted with Option 3, since it is not important to visit E0 to get to a few qualities of E1 and E2.
- Option-3: is helpful when the speculation is not add up to and the operations allude to either events and traits of E1(E2) or of E0, and in this way make qualifications among child and parent substances. For this situation, we can spare stockpiling contrasted with Option-1 as a result of the nonappearance of invalid esteems, in any case, there is an expansion of the quantity of gets to keep the events predictable.

There is a vital viewpoint that must be elucidated about the above. For the rebuilding of the speculations, the basic tallying of cases and accesses is not generally adequate for picking the most ideal choice. Given these elements, no doubt Option-3 would barely ever be appropriate, as it for the most part requires more gets to with a specific end goal to do operations on the information.

This rebuilding, notwithstanding, has the colossal preferred standpoint of producing elements with less qualities. As we might see, this converts into consistent and after that physical structures of little measurements for which a physical accesses permits the recovery of a more prominent measure of information (tuples) on the double. Accordingly, in a few basic cases, a more refined examination should be completed. This may take into account different elements, for example, the amount of information that can be recovered by methods for a solitary access to optional capacity.

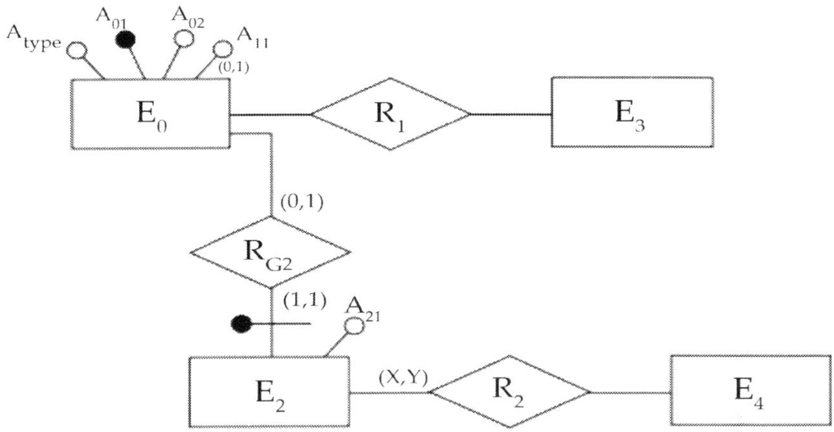

Figure 120 - Reconstructing of Previous Schemas.

(Source: http://pld.cs.luc.edu/database/)

The choices displayed are by all account not the only ones permitted, yet it is conceivable to do restructurings that join them. An illustration is given in Figure 120. For this situation, in light of contemplations like those talked about above, it was chosen to consolidate E0 and E1 and to leave the E2 entity isolated from the others. The ascribe A-type was added to recognize the events of E0 from those of E1.

At last, with respect to speculations on more than one level, breaking down a speculation at any given moment, beginning from the base of the whole chain of importance. Different designs are conceivable, which can be gotten by joining the essential restructurings on the different levels of the chain of importance.

Partitioning and Merging of Entities and Relations

Entities and relations of an E-R schema enhance the effectiveness of operations, utilizing the accompanying rule. Accesses are decreased by isolating properties of a similar idea that are gotten to by various operations and by consolidating traits of various ideas that are gotten to by similar operations. An indistinguishable criterion from those talked about for redundancies are legitimate in settling on a choice about this kind of rebuilding.

Partitioning of Entities

A case of entity partitioning is appeared in Figure 121.The EMPLOYEE entity is substituted by two elements, connected by a balanced relationship. One of them portrays individual data of a representative. Alternate depicts data about the work of a representative. This rebuilding is helpful if the operations that every now and again include the first entity requirements, for an employee, either personal data or employment data. This is a case of vertical partitioning of entities, as in the idea is sub-partitioned by its attributes. It is likewise conceivable, notwithstanding, to complete level dividing in which the sub-division takes a shot at the events of elements. For instance, there could be a few operations that relate just to the examiners and others that work just on the sales representatives. For this situation, as well, it could be helpful to parcel the EMPLOYEE substance into two particular elements, ANALYST and SALESPERSON having the same attributes as the first entity. Note that level partitioning relates to the presentation of chains of importance at the legitimate level.

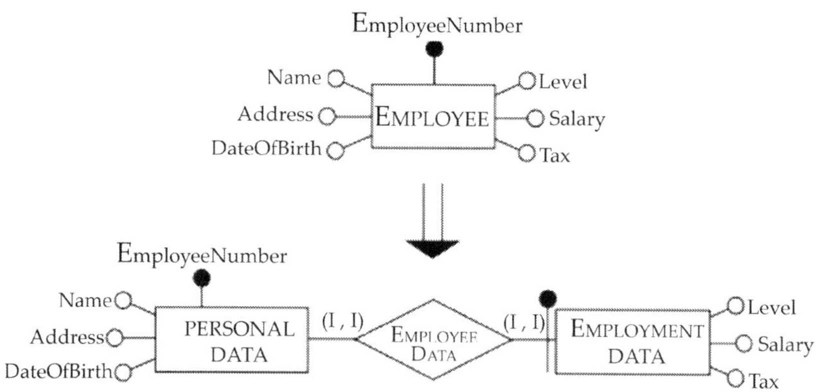

Figure 121 - Partitioning of Entities.

(Source: http://pld.cs.luc.edu/database/)

Even partitioning has the reaction of duplicating the connections in which the first substance took an interest. This marvel can efficiently affect the execution of the actual database. On the other hand, vertical apportioning creates substances with fewer traits. They can along these lines be converted into physical structures from which we can recover an incredible arrangement of information with solitary accesses.

Cancellation of Multi-Value Attributes

One sort of partitioning that is ought to be talked about, independently manages the cancellation of multi-values attributes. This rebuilding is important in light of the fact that, as with speculations, the relational model does not permit the immediate portrayal of these qualities.

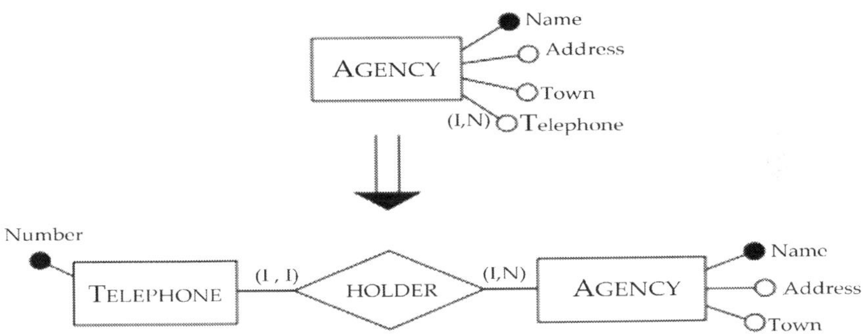

Figure 122 - Multi-Value Attributes.

(Source: http://pld.cs.luc.edu/database/)

The rebuilding required is very basic and is outlined in Figure 122. The AGENCY substance is isolated into 2 elements: an element having name and attributes as the first element, aside from the multi-esteemed characteristic Telephone, and another TELEPHONE substance, with the property Number. These elements are connected by a one-to-many relationship. Clearly, if the characteristic had additionally been discretionary, at that point the base cardinality for the AGENCY entity in the subsequent model would have been 0.

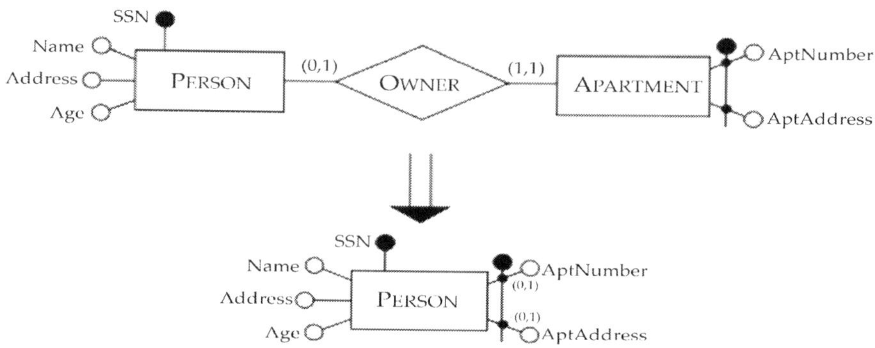

Figure 123 - Merging of the Entities.

(Source: http://pld.cs.luc.edu/database/)

Merging of Entities

Merging is the turnaround operation of partitioning. A case of merging of entities is appeared in Figure 123 in which the PERSON and APARTMENT entities, connected by an OWNER relationship, are converted into a solitary element having the qualities of both. This rebuilding can be proposed by the way that the most successive operations on the PERSON entity dependably require information identifying with the apartment that the person has. Along these lines, we wish to stay away from the accesses important to recover this information by methods for the OWNER relationship. A reaction of this re-building is the conceivable nearness of invalid esteems in light of the fact that, as indicated by the cardinalities, there are individuals who do not possess an apartment. Along these lines, there are no values for them for the APTADDRESS and APTNUMBER attributes.

Merging is for the most part done on coordinated connections, once in a while on one-to-many connections and barely ever on many-to-many relationship.

This is on the grounds that merging of entities connected by a one-to-many relationship or a many-to-many relationship creates redundancies. Specifically, it is simple to check that redundancies that show in non-key attributes of the entity, which takes an interest in the first association with a greatest cardinality equivalent to N.

Different sorts of partitioning and merging operations can also be linked to connections. This should be possible for two reasons. The first is a specific end goal to isolate events of a relationship that are continuously accessed independently. The second is to consolidate with a minimum of two connections between similar entities, into a solitary relationship, when their events are constantly accessed together. A case of partitioning of relationship is given in Figure 124 in which the present players of a ball group are recognized from past players.

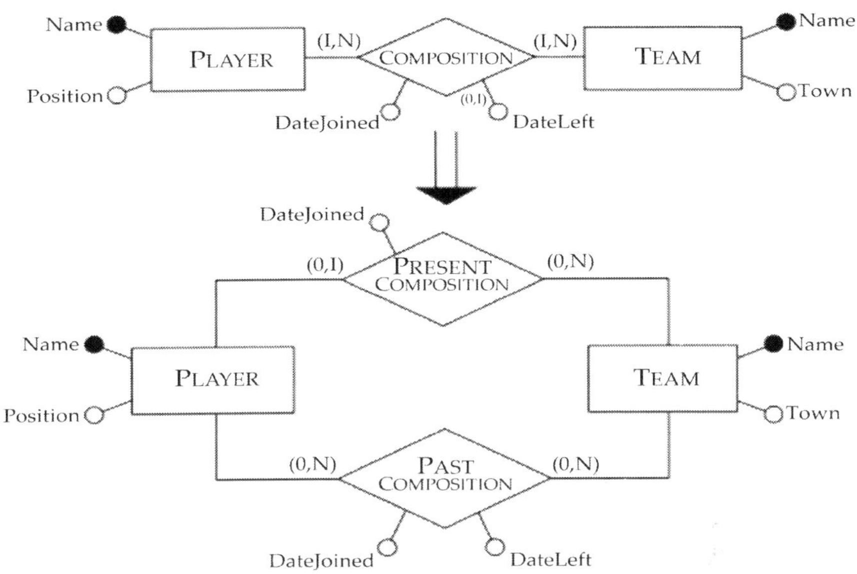

Figure 124 - Partitioning of Relationships.

(Source: http://pld.cs.luc.edu/database/)

We should specify that the choices about dividing and combining can be delayed until the point when of the physical outline stage. A considerable lot of the present database administration systems permit the particular of groups of intelligent structures, gathering of tables, completed at the physical level. Groups permit quick access to information circulated all through various intelligent structures.

Selection of Essential Identifiers

The decision of an identifier for every element is fundamental to the interpretation into the relational model, as a result of the significant part enters play in an esteem based information display. Besides, database administration systems require the determination of an essential key on which helper structures for quick access to information, known as records, are naturally built. In this way, where there are substances for which numerous identifiers (or actually none) have been determined, it is important to choose which characteristics constitute the essential identifier.

The criteria for this choice are as per the following.

- Attributes with invalid values cannot be assigned as identifiers. These qualities do not ensure access to every one of the events of the relating substance, as we called attention to while talking about keys for the relational model.
- One or few ascribes are desirable over many qualities. This guarantees the records are of restricted size, less capacity is required for the making of coherent connections among the different relations, and join operations are encouraged.
- For the same type of reason, an internal identifier with some attributes is more likable as compared to an external one. This is because external identifiers are translated in keys.
- An identifier that is used by real operations to access the occurrences of entities is more likable compared to others. These operations can actually be executed in a more efficient way, since they can take pros of the indices directly built from the DBMS.

TRANSLATION INTO THE RELATIONAL MODEL

The second step of logical model plan relates to an interpretation between diverse data models. Beginning from an E-R schema, a proportionate relational model is developed. By proportional, we mean a pattern able to do speaking to a similar data. As indicated by the rebuilding made on the E-R schema in the initial step of sensible outline, it is adequate to consider a streamlined rendition of the E-R model. In this form, a construction contains no speculations or multi-values attributes and has just essential identifiers.

We will manage the interpretation issue efficiently, starting with the crucial case, that of substances connected by many-to-many connections. This case exhibits the general guideline on which the entirety interpretation strategy is based.

Entities and Many-to-Many Relationship

Consider the diagram in Figure 125. Its characteristic interpretation into the relational model permits the accompanying:

Figure 125 – Many-to-Many Relationship.

(Source: Ramakrishnan, R. Database Management Systems.)

- for every entity, a connection with a similar name, having as attributes the same qualities as the entity and having its identifier as a key; and
- for the relationship, a connection with a similar name, having as attributes the qualities of the relationship and the identifiers of the entity included; these identifiers, taken together, frame the key of the relationship.

In the event that the first properties of entities or relationships are discretionary, the comparing properties of relations can expect invalid esteems.

The relational pattern got is in this manner as takes after:

EMPLOYEE(*Number* , Surname, Salary)

PROJECT(*Code* Code, Name, Budget)

PARTICIPATION(*Number* , *Code* , StartDate)

To make the significance of the construction clearer, it is useful to do a few renaming. For instance, for our situation we can illuminate the substance of the PARTICIPATION connection by characterizing it as takes after:

PARTICIPATION(*Employee*, *Project*, StartDate)

The space of the Employee property is an arrangement of employee numbers and that of the Project characteristic is an arrangement of project codes. There are referential requirements between these characteristics and the EMPLOYEE connection and the PROJECT connection individually.

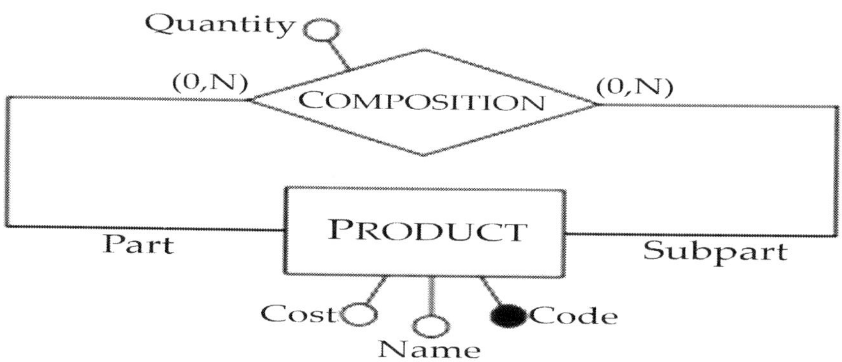

Figure 126 - E-R Recursive Relationship.

(Source: Ramakrishnan, R. Database Management Systems.)

Renaming is fundamental sometimes. For example, we have recursive relationship shown in Figure 126.

This construction is converted into two relations:

PRODUCT(*Code*, Name, Cost)

COMPOSITION(*Part*, *Subpart*, Quantity)

In this construction, both the properties Part and Subpart have item codes as space. There is in actuality a referential limitation between each of them and the PRODUCT connection.

The interpretation of a relationship with more than two elements is comparative to the interpretation of a parallel relationship. For instance, consider the outline with a ternary-relationship in Figure 127.

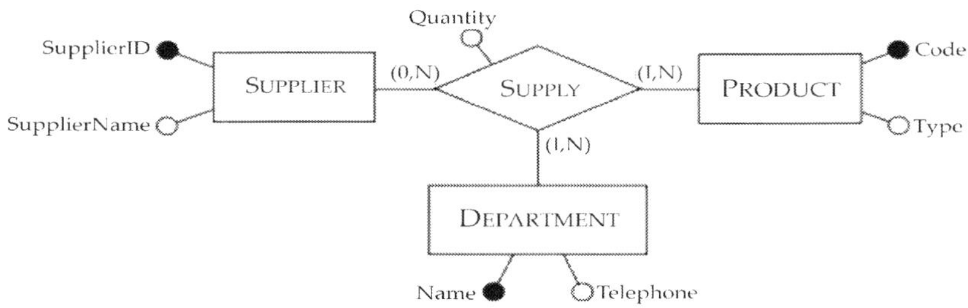

Figure 127 - E-R with Ternary-Telationship.

(Source: Ramakrishnan, R. Database Management Systems.)

This construction is converted into the accompanying four relations:

SUPPLIER(*SupplierID*, SupplierName)

PRODUCT(*Code*, Type)

DEPARTMENT(*Name*, Telephone)

SUPPLY(*Supplier*, *Product*, *Department*, Quantity)

There are referential limitations for the outline consequently acquired between the properties Supplier, Product, and Department of the SUPPLY connection and, separately, the SUPPLIER connection, the PRODUCT connection, and the DEPARTMENT connection. In this last kind of interpretation, we have to confirm whether the identifiers of the substances, taken together, do not constitute a key, rather, a repetitive super-key for the connection that speaks to the relationship of the E-R model. This can happen, for instance, on account of the pattern in Figure 127, if there is a sole provider who supplies an offered item to a given office.

Note that the cardinality is as yet substantial, since this provider can supply numerous items to this or different departments. For this situation, the way to the SUPPLY connection would be comprised of the characteristics Product and Department as it were, since, given a product and a department, the supplier is unambiguously decided.

One-to-Many Relationship

Consider the blueprint with a one-to-many relationship in Figure 128.

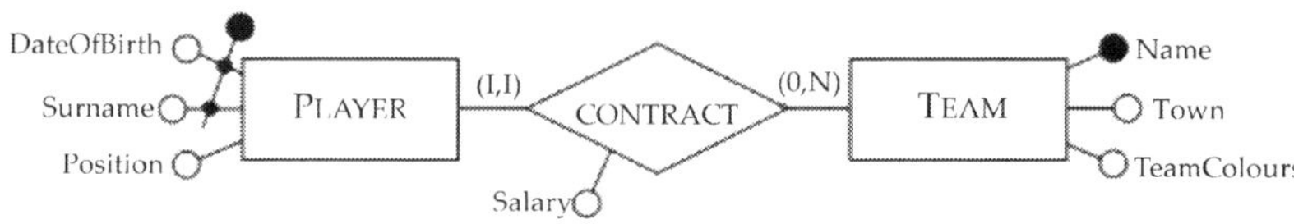

Figure 128 - E-R with One-to-Many Relationships.

(Source: Ramakrishnan, R. Database Management Systems.)

As per the control seen for one-to-many relationships, the interpretation of this mapping would be as per the following:

Note that in the CONTRACT connection, the key comprises of just the identifier of PLAYER in light of the fact that the cardinalities of the relationship disclose to us that every player has an agreement with a solitary group. Now, the relations PLAYER and CONTRACT have a similar key (the Surname and the DateOfBbirth of a player) and we can subsequently blend them in a solitary connection with no danger of repetition. This is on account that there is a coordinated correspondence between the particular examples. In this way, for the construction in Figure 128, the accompanying interpretation is ideal, in which the PLAYER connection speaks to both the element and the relationship of the first E-R schema:

PLAYER(*Surname* , *DateofBirth* , Position, Team, Salary)

TEAM(*Name* , Town, TeamColours)

In the schema, there is clearly the referential requirement between the trait Team of the PLAYER connection and the TEAM connection.

Note that the interest of the PLAYER entity is obligatory. In the event that it were discretionary (it is conceivable to have players who have no agreement with a group), both of the interpretations with three relations and with two relations would be legitimate. Regardless of the possibility that in the second interpretation we have fewer relations, it is in actuality conceivable to have invalid esteems in the PLAYER connection on the qualities Team and Salary. On the other hand, in the principal interpretation, this cannot occur.

In any case, when a substance takes an interest with a most extreme cardinality of one, we can spare a connection, as occurred in the interpretation of one-to-many twofold connections. The element that takes an interest in the association with most extreme cardinality of one, is converted into a connection that incorporates the identifiers of alternate elements engaged with the relationship (and also conceivable qualities of the relationship itself). There is, in this way, not anymore any need to speak to the first association with a different connection. As a case, accept that the PRODUCT substance took an interest in the relationship in Figure 127 with a base and most extreme cardinality of one. This implies that, for every product there is a sole supplier which supplies it and a sole department that is provided. At that point, the pattern is interpreted as the following.

SUPPLIER(*SupplierID* , SupplierName)

DEPARTMENT(*Name* ,Telephone)

PRODUCT(*Code* , Type, Supplier, Department, Quantity)

Here, there are referential requirements between the Supplier attribute of the PRODUCT connection and the SUPPLIER connection, and between the Department attribute of the PRODUCTs connection and the DEPARTMENT connection.

Entities with External Identifiers

Elements with external identifiers offer ascent to relations having keys that contain the identifier of the "recognizing" substances. Consider, for instance, the E-R schema appeared in Figure 129.

Figure 129 - E-R with External-Identifier.

(Source: Ramakrishnan, R. Database Management Systems.)

The comparing relational outline is as per the following:

STUDENT(<u>*RegistrationNumber*</u>, <u>*University*</u>, Surname, EnrolmentYear)

UNIVERSITY(<u>*Name*</u>, Town, Address)

in which there is a referential imperative between the characteristic University of the STUDENT connection and the UNIVERSITY connection.

As should be obvious, by speaking to the external identifier, we additionally speak to the connection between the two substances. Keep in mind that elements distinguished remotely, dependably partake in the association with a base and greatest cardinality of one. This sort of interpretation is substantial autonomously of the cardinality with which alternate elements take an interest in the relationship.

One-to-One Relationship

For balanced connections, there are for the most part numerous potential outcomes for interpretation. We will start with coordinated associations with compulsory cooperation for both the substances, for example, that in the mapping in Figure 130.

Figure 130 - E-R with One-to-One Relationships.

(Source: Ramakrishnan, R. Database Management Systems.)

There are two symmetrical and similarly legitimate potential outcomes for this kind of relationship.

With the referential requirement between the quality Department of the HEAD connection and the DEPARTMENT connection, or:

HEAD(<u>*Number*</u> , Name, Salary, Department, StartDate)

DEPARTMENT(<u>*Name*</u>, Telephone, Branch)

for which there is the referential requirement between the quality Head of the DEPARTMENT connection and the HEAD connection.

HEAD(<u>*Number*</u> , Name, Salary)

DEPARTMENT(<u>*Name*</u>, Telephone, Branch, Head, StartDate)

Since there is a coordinated correspondence between the events of the two entities, a further choice would appear to be conceivable in which we have a solitary connection containing every one of the characteristics in the mapping. This choice ought to be disposed of since we should not overlook that the construction that we are interpreting is the aftereffect of a procedure in which exact decisions were made with respect to combining and the apportioning of elements. This implies, if the rebuilt E-R diagram has two elements connected by a balanced connection, we thought that it was advantageous to keep the two ideas independent. It is in this manner not suitable to consolidate them amid the interpretation into the social model.

Let us consider the instance of a coordinated association with discretionary interest for one of the elements, for example, that in the pattern in Figure 131.

Figure 131 – One-to-One Relationship.

(Source: Ramakrishnan, R. Database Management Systems.)

For this situation we have one ideal alternative for which there is the referential requirement between the trait Head of the DEPARTMENT connection and the EMPLOYEE connection. This alternative is desirable over the other in which the relationship in the EMPLOYEE connection is referred through the name of the department, on the grounds that, for this trait, we could have invalid esteems.

EMPLOYEE(*Number* , Name, Salary)

DEPARTMENT(*Name*, Telephone, Branch, Head, StartDate)

At long last, consider the case in which both the entities have discretionary cooperation. For instance, consider the pattern in Figure 131, there can be departments with no head (and in this manner the base cardinality of the Department entity is equivalent to 0). For this situation, there is a further plausibility that considers three separate relations:

EMPLOYEE(*Number* , Name, Salary)

DEPARTMENT(*Name*, Telephone, Branch)

MANAGEMENT(*Head* , Department, StartDate)

The key of the MANAGEMENT connection could be the Department attribute also. Here, we have referential limitations between the Head and Department of the MANAGEMENT connection and the EMPLOYEE as well as of the DEPARTMENT relations, separately. This arrangement has leeway of failing to have invalid esteems on the qualities that actualize the relationship. Then again, we require an additional connection, with a subsequent increment of the multifaceted nature of the database.

Along these lines, the three-connection arrangement is to be viewed as just if the number of events of the relationship is low contrasted with the events of the substances engaged with the relationship. For this situation, there is the preferred standpoint of maintaining a strategic distance from the nearness of numerous invalid esteems.

Translation of an Unpredictable Pattern

To perceive how to continue in a mind boggling case, we will do a total case of an interpretation in light of the outline appeared in Figure 132.

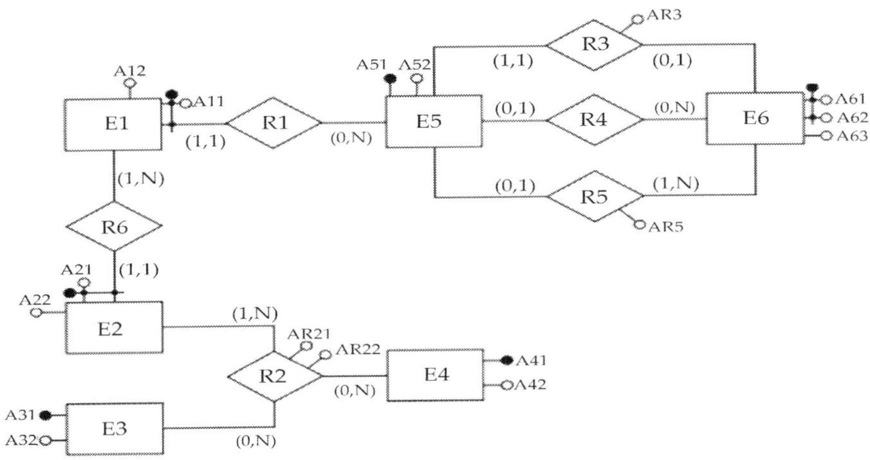

Figure 132 - Translation for E-R Schema.

(Source: http://www.learndb.com/databases/how-to-convert-er-diagram-to-relational-database)

In the principal stage, we make an interpretation of every substance into a connection. The interpretation of the substances with inward identifiers is quick:

$$E3(\underline{A31}, A32)$$

$$E4(\underline{A41}, A42)$$

$$E5(\underline{A51}\ A51, A52)$$

$$E6(A61, \underline{A62}, A63)$$

Presently we interpret the substances with outer identifiers. We acquire this relations found below.

$$E1(\underline{A11}, \underline{A51}, A12)$$

$$E2(\underline{A21}, \underline{A11}, \underline{A51}, A22)$$

We should note how E-2 takes the property A11 and (transitively) the quality A51, which, together with A21, distinguishes E-1. Some referential imperatives are characterized for the relations delivered (for instance, there is a referential limitation between the characteristic A51 of E1 and E5).

We now proceed onward to the interpretation of connections. Connections R1 and R6 have just been deciphered in light of the outside ID of E2 and E1 separately. We expect we have chosen to get a base number of relations in the last blueprint and we will attempt consequently to consolidate relations where conceivable. We get the accompanying changes to be conveyed out on the underlying relations:

- so as to interpret R3, we present, with fitting renaming, the characteristics that distinguish E6, among those of E5; and also the trait AR3 of R3. In this way, we present A61R3, A62R3 and AR3 in E5;
- comparably for R4; we present A61R4 and A62R4 into E5;
- comparably for R5; we present A61R5; A62R5 and AR5 into E5.

Note that the renaming is actually basic and fundamental, with a specific end goal to have the capacity to recognize between the employments of similar credits to speak to various connections (for instance, A61R3, which speaks to R3, and A61R4, which speaks to R4).

We make an interpretation of the main many-to-many relationship:

$$\mathrm{R2}(\underline{A21}, \underline{A11}\ , \underline{A51}, \underline{A31}, \underline{A41}, \mathrm{AR21}, \mathrm{AR22})$$

The Relational model got is along these lines as takes after:

$$\mathrm{E1}(\underline{A11}, \underline{A51}, \underline{A11}\ , \underline{A51}, \mathrm{A12})$$

$$\mathrm{E2}(\underline{A21}, \underline{A11}, \underline{A51}, \mathrm{A22})$$

$$\mathrm{E3}(\underline{A31}, \mathrm{A32})$$

$$\mathrm{E4}(\underline{A41}, \mathrm{A42})$$

$\mathrm{E5}(\underline{A51}$, A52, A61R3, A62R3, AR3, A61R4, A62R4, A61R5, A62R5, AR5)

$$\mathrm{E6}(\underline{A61}, \underline{A62}, \mathrm{A63})$$

R2(<u>*A*21</u>, <u>*A*11</u> , <u>*A*51</u>, <u>*A*31</u>, <u>*A*41</u>, AR21, AR22)

Give us a chance to take note of that we have gotten relations (E2 and R2) with keys made out of many characteristics. In such cases one could choose to present straightforward keys (codes) either at this stage or prior in the rebuilding venture.

Summary Tables

The interpretations we have seen are outlined in Figure 133 and Figure 134.

For each sort of arrangement of E-R schema, a depiction of the case and the conceivable interpretations are provided.

In these tables, the images X and Y demonstrate any of the permitted cardinalities. The reference bullets demonstrate the likelihood of having invalid esteems on the related traits and the broken underline shows an option key to the one demonstrated by a strong underline.

Documentation of Logical Schemas

Similarly as with conceptual design, the aftereffect of logical schema does not comprise only of a straightforward database-schema but in addition of the documentation related with it. To start with, a significant part of the documentation of the conceptual design created in the legitimate outline stage can be acquired by the intelligent composition. In specific, if the names of the ideas of the E-R mapping are reused to develop the relational pattern, the business rules characterized already can likewise be utilized for actual documentation of the last one. This type of documentation must be finished with the referential limitations presented by the interpretation.

In this unique situation, we can receive a straightforward graphical documentation for the portrayal of the two relations and the referential-limitations existing between the different relations. A case of this documentation is given in Figure 135. In this portrayal keys of relations show up in striking face, bolts depict referential limitations and the nearness of bullets on the properties signifies the likelihood of having invalid esteems on them.

Type	Initial Schema	Possible translation
Binary many-to many relationship	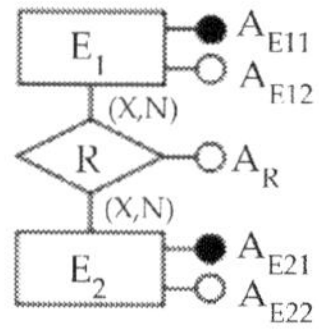	$E_1(\underline{A_{E11}} \cdot A_{E12})$ $E_2(\underline{A_{E21}} \cdot A_{E22})$ $R(\underline{A_{E11}}, \underline{A_{E21}}, A_R)$
Ternary many-to many relationship	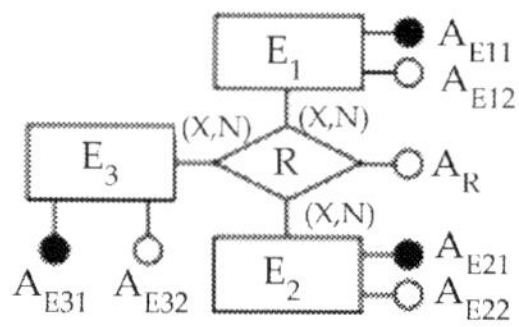	$E_1(\underline{A_{E11}} \cdot A_{E12})$ $E_2(\underline{A_{E21}}, A_{E22})$ $E_3(\underline{A_{E31}}, A_{E32})$ $R(\underline{A_{E11}}, \underline{A_{E21}}, \underline{A_{E31}}, A_R)$
One-to-many relationship with mandatory participation	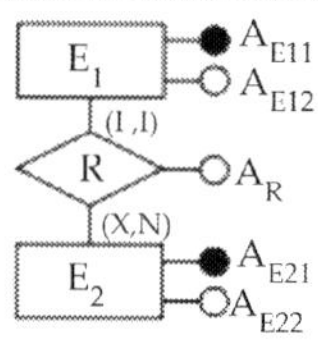	$E_1(\underline{A_{E11}}, A_{E12}, A_{E21}, A_R)$ $E_2(\underline{A_{E21}}, A_{E22})$
One-to-many relationship with optional participation	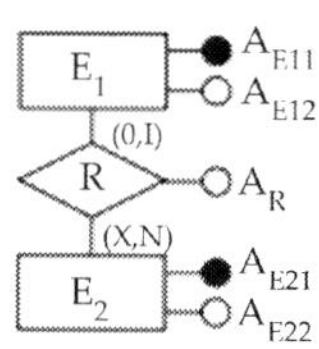	$E_1(\underline{A_{E11}}, A_{E12})$ $E_2(\underline{A_{E21}}, A_{E22})$ $R(\underline{A_{E11}}, \underline{A_{E21}}\ A_R)$ Alternatively: $E_1(\underline{A_{E11}}, A_{E21}, A^*_{E21}, A^*_R)$ $E_2(\underline{A_{E21}}, A_{E22})$
Relationship with external identifiers	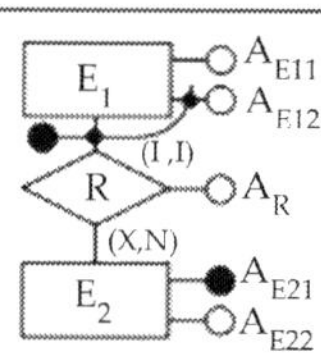	$E_1(\underline{A_{E12}}, \underline{A_{E21}}, A_{E11}, A_R)$ $E_2(\underline{A_{E21}}, A_{E22})$

Figure 133 - Translations of E-R Model.

(Source: http://www.learndb.com/databases)

Type	Initial Schema	Possible translation
One-to-one relationship with mandatory participation for both entities	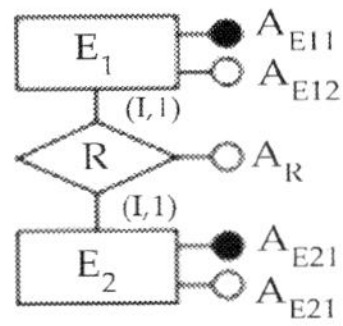	$E_1(A_{E11} . A_{E12}\ \underline{A_{E21}} . A_R)$ $E_2(\underline{A_{E21}} . A_{E22})$ Alternatively: $E_2(A_{E21} . A_{E22}\ \underline{A_{E11}}\ A_R)$ $E_1(A_{E11} . A_{E12})$
One-to-one relationship with optional participation for one entity	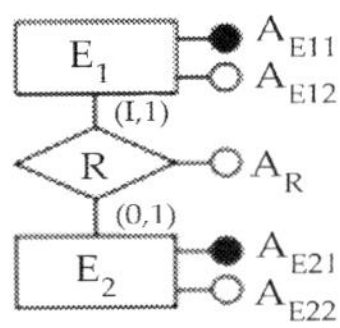	$E_1(\underline{A_{E11}} . A_{E12}\ \underline{A_{E21}} . A_R)$ $E_2(\underline{A_{E21}} . A_{E22})$
One-to-one relationship with optional participation for both entities	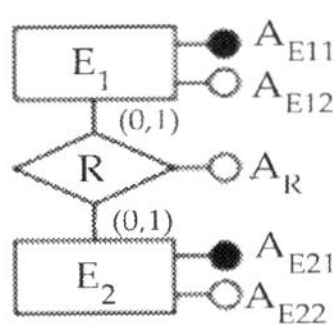	$E_1(\underline{A_{E11}} . A_{E21})$ $E_2(\underline{A_{E21}} . A_{E22} . A^{*}_{E11} . A^{*}_{R})$ Alternatively: $E_1(\underline{A_{E11}} . A_{E12} . A^{*}_{E21} . A^{*}_{R})$ $E_2(\underline{A_{E21}} . A_{E22})$ Alternatively: $E_1(\underline{A_{E11}} . A_{E12})$ $E_2(\underline{A_{E21}} . A_{E22})$ $R(\underline{A_{E11}} . \underline{A_{E21}} . A_R)$

Figure 134 - Translations of E-R Model.

(Source: http://www.learndb.com/databases)

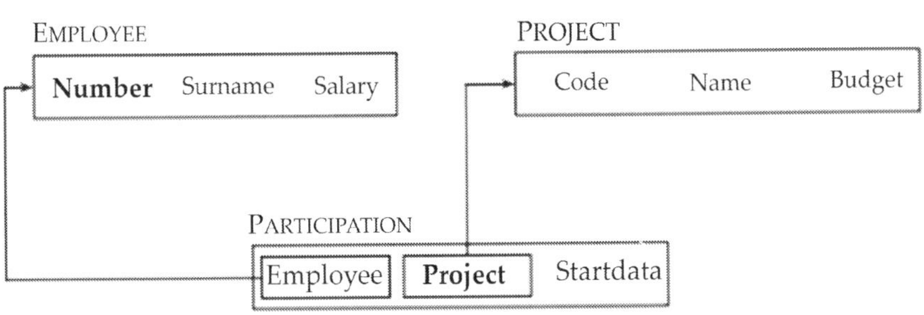

Figure 135 - Graphical Representation of the Translation.

(Source: http://www.learndb.com/databases)

Along these lines, we can monitor the connections of the first E-R model. Subsequently, for the situation in the illustration, the activities in which the representatives take an interest can be recovered by methods for the PARTICIPATION connection.

Another case of the utilization of this documentation is given in Figure 136.

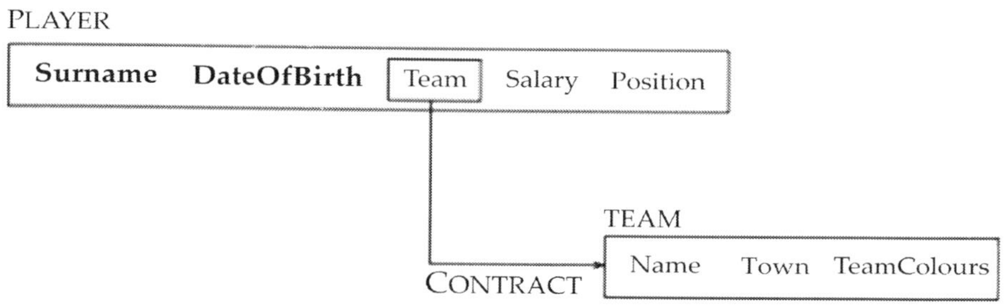

Figure 136 - Graphical representation of the Ttranslation.

(Source: http://www.learndb.com/databases)

Note that this technique likewise enables us to speak to expressly the connections of the first E -R model, to which in the equal relational outline, no connection compares (the CONTRACT relationship in the case being referred to).

As a last case, Figure 137 demonstrates the portrayal of the relational schema. Presently, the legitimate connections between the different relations can be effortlessly recognized.

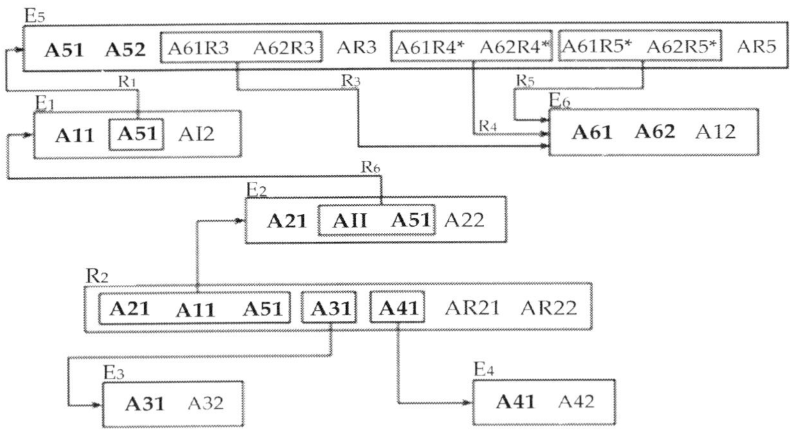

Figure 137 - Graphical Representation of a Relational Schema.

(Source: http://www.learndb.com/databases)

A CASE OF LOGICAL DESIGN

Give us a chance to come back to the case in the former part with respect to the preparation-training of the organization. The applied composition is indicated once more, for comfort, in Figure 138.

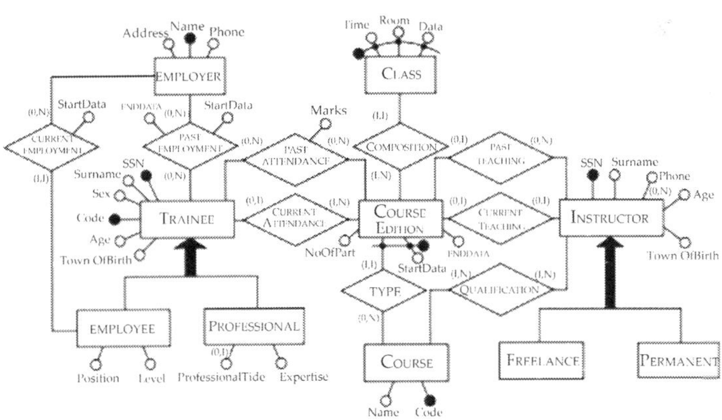

Figure 138 - E-R Schema for Training Company.

(Source: http://www.learndb.com/databases)

The accompanying operations were moved toward the information portrayed by this pattern:

- operation-1: embed another trainee demonstrating all his or her information;
- operation-2: relegate a trainee to a course edition;
- operation-3: embed another instructor showing all his or her information and the courses he or she is qualified to handle;
- operation-4: dole out a qualified instructor to a course edition;
- operation-5: show all the data on the past course editions with name, class time, and number of trainees;
- operation-6: show every course offered, with data on the instructors who are qualified for these;
- operation-7: for every instructor, discover the trainees for every course he or she had handled;
- operation-8: do a factual examination of the considerable number of trainees with all the data on them, based on the course editions they have attended and the corresponding marks received.

Restructuring Phase

We will now do the different rebuilding undertakings. The different changes are indicated together in the last mapping in Fig 139.

Table of volumes		
Concept	**Type**	**Volume**
Class	E	8000
ClassEdition	E	1000
Course	E	200
Instructor	E	300
Freelance	E	250
Permanent	E	50
Trainee	E	5000
Employee	E	4000

Professional	E	1000
Employer	E	8000
PastAttendance	R	10000
CurrentAttendance	R	500
Composition	R	8000
Type	R	1000
PastTeaching	R	900
CurrentTeaching	R	100
Qualification	R	500
CurrentEmployment	R	4000
PastEmployment	R	10000

Table of operations		
Operation	**Type**	**Frequency**
Op 1	1	40 per day
Op 2	1	50 per day
Op 3	1	2 per day
Op 4	1	15 per day
Op 5	1	10 per day
Op 6	1	20 per day
Op7	1	5 per day
Op 8	B	10 per month

Figure 139 - Tables of the Volumes and Operations.

Investigation of redundancies - There is just a single repetitive bit of information in the pattern: the trait NumberOfParticipants in COURSEEDITION, which can be gotten from the connections CURRENTATTENDANCE and PASTATTENDANCE.

The capacity necessity is 4×1000=4000 bytes, having accepted that 4bytes are fundamental for each event of COURSE-EDITION to store the numbers of members. The operations required with this data are: 2, 5, 8. The remainder of these can be forgotten in light of the fact that it manages a rare operation that is completed in group mode. We will in this manner assess the cost of operations 2 and 5 in the instances of the presence or non-presence of excess information. We can conclude from the table of volumes that every version of the course has, all things considered, 8 classes and 10 members. From this information we can without much of a stretch infer the get to tables.

Accesses with redundancy			
Operation 2			
Concept	**Cnstr**	**Acc.**	**Type**
Trainee	E	1	R
CurrentAtt'nce	R	1	W
CourseEdition	E	1	R
CourseEdition	E	1	W

Operation 5			
Concept	**Cnstr**	**Acc.**	**Type**
CourseEdition	E	1	R
Type	R	1	R
Course	E	1	R
Compostion	R	8	R
Class	E	8	R

Accesses without redundancy			
Operation 2			
Concept	**Cnstr**	**Acc.**	**Type**
Trainee	E	1	R
CurrentAtt'nce	R	1	W

Operation 5			
Concept	**Cnstr**	**Acc.**	**Type**
CourseEdition	E	1	R
Type	R	1	R

Course	E	1	R
Compostion	R	8	R
Class	E	8	R
PastAttendance	R	10	R

Figure 140 - *2 Access Table.*

(Source: http://www.databasejournal.com/)

From the given to tables we acquire:

- with repetition: for operation 2 we have 2×50=100 read accesses and the same number of again in compose gets to every day, meanwhile, for operation-5, we have 19×10=190 read accesses every day for an aggregate of 490 accesses every day (having given twofold weight to the compose gets to);
- without repetition: for operation 2 we have 50 perused accesses every day furthermore, the same number of again in compose gets to every day, while, for operation 5, we have 29×10=290 read accesses every day, for an aggregate of 440 accesses every day (having already given twofold weight to compose accesses).

Subsequently, when the excess is available, we have disservices both in terms of capacity and get to time. We will in this way erase the quality NumberOfParticipants from the element COURSEEDITION.

Removing Generalizations

There are two generalizations in the pattern: that identifying with the instructors and that identifying with the trainees. For the instructors, it can be noticed that the important operations: 3; 4; 6 and also 7, see no difference amongst independent instructors and those utilized on a changeless premise by the organization. Moreover, the relating elements have no particular characteristics for them. In this manner, we choose to erase the type elements of the speculation and add a credit 'Type' to the 'INSTRUCTOR' element. This trait has an area made up of the images F (for independent) also, P (for changeless).

For the trainees, the operations including this information (operations 1, 2 and also 8) have no significant effect between the different sorts of event. We can see from the composition that both professionals and employees have particular attributes. We should, subsequently, leave the EMPLOYEE and PROFESSIONAL entities, including two one-tone connections between these entities and the TRAINEE entity. Along these lines, we can abstain from having traits with conceivable invalid esteems on the parent element of the speculation and we can diminish the measurement of the relations.

Partitioning and Merging Entities

From the examination of information and operations, numerous potential restructurings of this sort can be recognized. The first one identifies with the COURSEEDITION entity. We can see that operation-5 relates just to the past versions and that the connections PASTTEACHING and PASTATTENDANCE allude only to these versions of the course. In this manner, keeping in mind the end goal to make the above operation more productive, we could break down the substance on a level plane to recognize the present versions from the previous ones. The burden of this decision is that the connections COMPOSITION and TYPE would be copied. Moreover, operations 7 and also 8 do not make incredible qualifications between current versions and past ones and would be more costly, on the grounds that they would expect visits to two particular elements.

Along these lines, we would not partition this substance. Two other conceivable restructurings that we could consider are the merging of the connections PASTTEACHING and PRESENTTEACHING and the comparative connections PASTATTENDANCE and PRESENTATTENDANCE. In the two cases, we are managing two comparative ideas between which a few operations make no distinction (7 and also 8). The merging of these connections would deliver another preferred standpoint: it would never again be important to exchange events starting with one relationship then onto the next toward the finish of a course edition. A negative factor is the presence of the trait Mark, which does not have any significant bearing to the current edition and could accordingly create invalid values. For the rest, the table of volumes reveals to us that the assessed number of events of the

CURRENTATTENDANCE relationship is actually 500. In this manner, assuming that we require 4 bytes to store the imprints, the misuse of capacity would be just 2 K-bytes.

We can choose in this way to combine the two sets of connections as portrayed in Figure 141. We should include a limitation that is not expressible by the diagram, which requires that an instructor cannot show more than one course edition in any one period. Also, a trainee cannot go to more than one course edition at a specific time.

We have to evacuate the multi-value property Telephone from the INSTRUCTOR entity. To do this we should present another entity TELEPHONE connected by a one to many association with the INSTRUCTOR entity, from which the characteristic will be expelled.

It is intriguing to take note of that some choices made in this stage turn around, in some kind of way, choices made during the applied design stage. This is definitely not astonishing in any case: the point of applied design is only to speak to the necessities in the most ideal path conceivable, without considering the effectiveness of the application. In coherent plan we should rather attempt to upgrade the execution and rethinking prior choices are unavoidable.

Choice of essential identifiers - Only the TRAINEE entity presents 2 identifiers: the SSN and the code. It is far desirable to choose the second. An SSN can require a few bytes while the code, which serves to recognize 5000 events requires close to 2 bytes.

There is even another even thought to be made with respect to identifiers, which has to do with the COURSEEDITION entity. This entity is recognized by the StartDate attribute and by the COURSE entity. This gives a composite identifier that, in a relational portrayal, must be utilized to execute two connections (ATTENDANCE and TEACHING). We can see that each one of the course has a specific code and that the normal number of course edition is five. This implies that it is adequate to add a little number to the course code to have the identifier for the course edition. This operation can be done proficiently and precisely amid the production of another edition. This is a case of investigation and rebuilding that is not in any of the general classes we have seen, however in practice can be experienced.

This is the finish of the rebuilding period of the first E-R mapping.

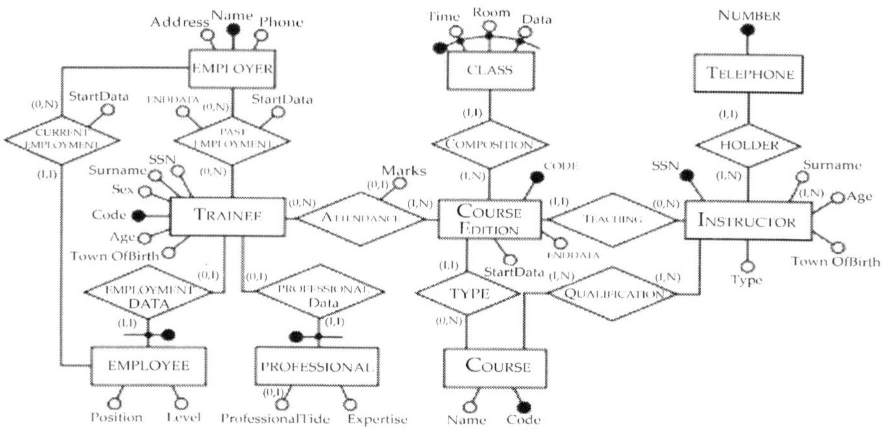

Figure 141 - 3 The E-R Schema of Fig. 138 After the Restructuring Phase.

(Source: http://www.databasejournal.com/)

Translation into the Relational Model

By following the interpretation methods depicted in this part, the E-R model in Figure 141 can be converted into the accompanying relational model. The sensible model will normally be finished by a help report that depicts, in addition to other things, all the referential imperatives that exist between the different relations.

COURSEEDITION (*Code* , StartDate, EndDate, Course, Instructor)

CLASS(*Time* , *Room* , *Date* , Edition)

INSTRUCTOR(*SSN* , Surname, Age, TownOfBirth, Type)

TELEPHONE(*Number* , Instructor)

COURSE(*Code* , Name)

QUALIFICATION (*Course* , *Instructor*)

TRAINEE (*Code* , SSN, Surname, Age, TownOfBirth, Sex)

ATTENDANCE(*Trainee*, *Edition*, Marks*)

EMPLOYER(*Name*, Address, Telephone)

PASTEMPLOYMENT(*Trainee*, *Employer*, StartDate, EndDate)

PROFESSIONAL(*Trainee*, Expertise, ProfessionalTitle*)

EMPLOYEE(*Trainee*, Level, Position, Employer, StartDate)

CASE TOOLS

The logical design phase is generally held by all the CASE tools for database advancement available. Specifically, since the interpretation to the relational model depends on exact criteria, it is done by these frameworks naturally. Then again, the rebuilding step, which goes before the real interpretation, is hard to robotize and the different items give practically no help to it. For instance, a few systems consequently decipher every one of the speculations as indicated by only one of the techniques portrayed before. We have seen, in any case, that the rebuilding of an E-R model is a crucial action of the plan for an imperative reason. To be specific, it can give answers for productivity issues that ought to be settled before completing the interpretation and that are most certainly not important to applied plan. The creator is ought to take care and to handle this perspective without putting excessively certainty into the instrument accessible.

A translation of the design using the CASE tools is presented in the below. Note how the many-to-many connection among EMPLOYEE and PROJECT has been converted into a connection. Likewise, take note of how new attributes have been added to the relations starting from entities with one-to-many and balanced connections. In the figure, the SQL-code moreover shows up, produced naturally by the system. It actually enables the planner to characterize the database utilizing a particular database administration system. A few of these systems permit the coordinate association with a DBMS and can develop the comparing database consequently. Different systems give devices the chance to convey out the turnaround operation: remaking a conceptual model in view of an existing relational model. This operation is called 'reverse engineering' and is especially valuable for the examination of a heritage system that is maybe situated towards a movement to another database administration system.

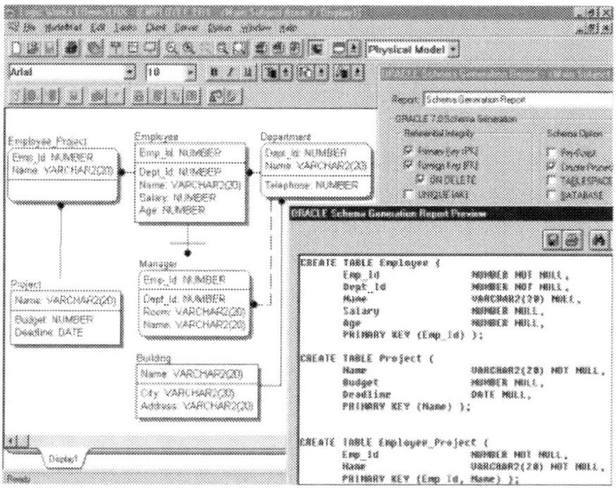

Figure 142 - CASE Tool.

(Source: http://www.databasejournal.com/)

CHAPTER 9

NORMALIZATION

In this section, we talk about a few properties, usually known as normal forms, which we can use to assess the nature of a relational database. We will see that when a relation does not fulfil a normal form, it presents redundancies and produces undesirable conduct in an update operation. This rule can be utilized to do quality examination on relational databases thus constitutes a valuable instrument for database design. For the designs that do not fulfil a normal form, we can apply a system known as normalization. Normalization permits the non-normalized tables to be changed into new patterns for which the fulfilment of a normal form is ensured. There are two critical focuses to be illuminated. To start with, the outline strategies found in the previous sections typically enable us to get blueprints that fulfil a normal form. In this structure, normalization hypothesis constitutes a valuable confirmation instrument, which shows corrections. Second, normalization hypothesis has been created with regards to the relational model and thus, it gives means and methods to the examination of the results of coherent outline. We will see, that the same strategies can be utilized, with minor varieties, on Entity-Relationship models.

We will manage this subject in stages, first talking about the issues (redundancies and inconsistencies) that can be confirmed in a relational construction, and at that point determining methodical procedures for examination and normalization.

REDUNDANCIES AND INCONSISTENCIES

We will utilize a case to outline the fundamental standards. Consider the relation in Figure 143. The key is comprised of the Employee and Project attributes. We can likewise effortlessly check that the connection fulfils the accompanying properties:

Employee	Salary	Project	Budget	Function
Brown	20	Mars	2	technician
Green	35	Jupiter	15	designer
Green	35	Venus	15	designer
Hoskins	55	Venus	15	manager
Hoskins	55	Jupiter	15	consultant
Hoskins	55	Mars	2	consultant
Moore	48	Mars	2	manager
Moore	48	Venus	15	designer
Kemp	48	Venus	15	designer
Kemp	48	Jupiter	15	manager

Figure 143 - Relationship with Different Anomalies.

(Source: http://pld.cs.luc.edu/database/)

- The salary of every employee is one of a kind and depends just on the employee. It is independent of the project on which the employee is assigned.
- The budget of each project is unique and depends just on the project. It is independent of the employees who are assigned to it.

These actualities have certain results for the substance of the connection and for the operations that can be completed on it. We will restrict our discourse to the main property, leaving the investigation of the second as an activity.

- The value of the salary of every employee is rehashed in all the tuples identifying with it. Hence, there is a repetition. If, for instance, an employee partakes in 20 projects, his or her salary will be repeated 20 times.

- If the salary of an employee changes, we need to adjust the salary in the entire relating tuples. This badly arranged process, which makes it important to do numerous adjustments at the same time, is known as update anomaly.
- If an employee quits with every one of the projects but does not leave the organization, all the relating tuples are deleted thus even the fundamental data such as name and salary that were not identified with projects are lost. This would expect us to keep a tuple with an invalid value for the quality Project, since Project is a piece of the key, this is not permitted. This issue is known as the delete anomaly.
- Similarly, on the off chance that we have data on another employee, we cannot insert it until the point that the employee is assigned to a project. This is known as the insert anomaly.

An instinctive clarification for the nearness of these undesirable wonders can be as per the following. We have utilized a solitary connection to speak to things of data of various sorts. Specifically, in this connection the following are spoken to: employees with their salaries, projects with their budgets, and cooperation of the employees in the projects they are assigned.

- Items of data regularly should be rehashed, that is, to show up in diverse tuples.
- If some data is rehashed repetitively, the updating must be rehashed for every event of the rehashed information. Presently, the relational languages, for example, SQL, all permit the determination of different updates utilizing a solitary order. In any case, this is just from the actual viewpoint for the issue from the perspective of the software engineer and not from that of the framework. The reason is that all the tuples engaged with the update must be altered and it is important to get to every last one of them physically.
- Problems can happen with the deletion of a tuple for the situation where only one thing of data it communicates is not any more substantial. The deletion can cause the cancellation of all the data being referred to, even the segment that is as yet legitimate.

- The insertion of solitary piece of data in a connection is impractical without values for the whole arrangement.

FUNCTIONAL CONDITIONS

To methodically think about the standards presented above, it is important to utilize a particular idea, practical reliance. This is a specific uprightness requirement for the relational model, which as the name recommends, portrays useful connections among the traits of a connection. Give us a chance to take a gander at the connection in Figure 143. We have actually seen that the salary of every employee is exceptional and in this manner each time a specific employee shows up in a tuple, the estimation of his or her salary dependably continues as before. We would thus be able to say that the value of the Salary attribute practically relies upon the estimation of the Employee attribute. A capacity exists with each Employee value in the connection, a solitary incentive for the Salary attribute. A comparable contention can be made for the connection between the Project and Budget attributes in light of the fact that the value of the budget practically relies upon the value of the project.

We can formalize this standard as follows. Given a connection r on a composition R(X) and two non-purge subsets Y and Z of the characteristics X, we say that there is an utilitarian reliance on r amongst Y and Z, if, for each combine of tuples t1 and t2 of r having similar esteems on the traits Y, t1 and t2 moreover have similar estimations of the Z characteristics.

A utilitarian reliance between the characteristics Y and Z is for the most part shown by the documentation $Y \rightarrow Z$ and, as with other uprightness requirements, is related with a construction: a legitimate connection on that outline needs to fulfil this utilitarian reliance. Given a utilitarian reliance Y to Z, we will call the Y the left hand-side and that Z the right-hand side of the reliance.

$$\text{Employee} \rightarrow \text{Salary}$$

$$\text{Project} \rightarrow \text{Budget}$$

Coming back to our case, we would thus be able to state that on the connection in Figure 143 there are the practical conditions:

$$\text{Employee Project} \rightarrow \text{Project}$$

There are a few perceptions to be made on practical conditions. In light of the given definition, we can express that, in our connection, this utilitarian reliance is additionally fulfilled: two tuples with similar values on each of the Employee and Project attributes have a similar incentive on the characteristic Project, which is 1 of them.

It is a "paltry" practical reliance since it affirms a self-evident property of the connection. Useful conditions ought to rather be utilized to portray critical properties of the app that we are speaking to. We at that point say that a practical reliance Y → Z is non-paltry if no quality in Z shows up among the traits of Y.

Note that a "half and half" practical reliance of the sort

Project → Project Budget

which contains an insignificant property (a venture relies upon itself), can be made non-insignificant by erasing from the correct hand side of the reliance, all the characteristics that likewise show up in the left-hand side. It is simple, actually, to show that if the useful reliance Y → Z is legitimate, at that point the practical reliance Y → W is likewise legitimate, where the W is a subset of the Z. From here on, we will allude practical conditions that are non-paltry, frequently discarding the modifier for curtness. A moment perception on practical conditions concerns their connection with the key limitation. On the off chance that we take a key K of a connection r, we can without much of a stretch confirm that there is a utilitarian reliance amongst K and some other property of r. This is on the grounds that, by the meaning of key requirement, there cannot exist two tuples with similar esteems on K. Alluding to our case, we have said that the traits Employee and Project shape a key. We would then be able to insist that, for case, the practical reliance Employee Project → Function is substantial. In specific, there will be a practical reliance between the key of a connection and the various characteristics of the pattern of the connection. For our situation we have:

Employee Project → Salary Budget Function

We can consequently finish up by saying that the practical reliance limitation sums up the key imperative. More exactly, we can state that a useful reliance Y → Z on a diagram R(X) savages into the key imperative if the union of the Y and the Z is equivalent to X. For this situation, Y is a super-key for the R(X) outline.

BOYCE–CODD NORMAL FORM

Definition of Boyce–Codd Normal Forma

Let us begin by saying that, in our case, the two properties causing anomalies relate precisely to qualities engaged with practical conditions.

- The property 'the salary of every employee is special and depends just on the employee, independent of the project in which he or she is assigned can be formalized by methods for the utilitarian reliance 'Employee → Salary'.
- The property 'the budget of each project is unique and depends just on the project, independent of the employees who are assigned in it' compares to the useful reliance 'Project → Budget'.

Moreover, it is fitting to take note of that the Function trait demonstrates, for each one of the tuples, the role of the worker in the venture. This part is remarkable, for every worker extend match. We can demonstrate this property as well utilizing a utilitarian reliance:

- The property 'in each project, each of the employee included is given a single function' relates to the useful reliance 'Employee Project → Function'. As we have said in the past segment, this is additionally an outcome of the way that the Employee and Project attributes frame the key of the connection.

We talked about how the initial two properties (and along these lines the relating practical conditions) create undesirable redundancies and inconsistencies. The third reliance is extraordinary. It never creates redundancies in light of the fact that, having Employee and Project as keys, the connection cannot contain two tuples with similar estimations of these properties (and along these lines of the Function property). Additionally, from a theoretical perspective, we can state that it cannot produce inconsistencies, in light of the fact that every employee has a project and each project has a budget; hence, hence for each employee extend match we can have remarkable esteems for the various qualities of the connection. Such esteems would not be accessible. In these cases, since they are not some portion of the key, we would just supplant them with invalid esteems with no issue. We would thus be able to infer that the conditions:

$$\text{Employee} \rightarrow \text{Salary}$$

$$\text{Project} \rightarrow \text{Budget}$$

cause anomalies, while the reliance

$$\text{Employee Project} \rightarrow \text{Function}$$

does not. The distinction, as we have specified, is that Employee Project is a super-key of the connection. All the thinking that we have created with reference to this particular illustration is broader. Without a doubt, redundancies and inconsistencies are caused by the practical conditions 'X → Y' that permit the actual presence of many equivalent tuples on the qualities in X. That is from the practical conditions 'X → Y' with the end goal that X does not contain the key. We will actually formalize this thought by presenting the idea of Boyce–Codd normal form, which takes the name from its creators. A connection r is in Boyce-Codd normal form if for each non-trivial functional dependency 'X → Y' exists, 'X' contains the key K of r; X is a super-key for r.

Anomalies and redundancies, as talked about above, do not show up in databases with relations in Boyce–Codd normal form, in light of the fact that the free snippets of data are discrete, one for every connection.

Decomposition into Boyce–Codd Normal Form

Given a connection that does not fulfil Boyce–Codd normal form, we can frequently supplant it with at least one standardized relations utilizing a procedure called normaliation. This procedure depends on a straightforward basis: if a connection speaks to some true ideas, at that point it is deteriorated into littler relations, one for every idea.

Employee	Salary
Brown	20
Green	35
Hoskins	55
Moore	48
Kemp	48

Project	Budget
Mars	2
Jupiter	15
Venus	15

Employee	Project	Function
Brown	Mars	technician
Green	Jupiter	designer
Green	Venus	designer
Hoskins	Venus	manager
Hoskins	Jupiter	consultant
Hoskins	Mars	consultant
Moore	Mars	manager
Moore	Venus	designer
Kemp	Venus	designer
Kemp	Jupiter	manager

Figure 144 - Decomposition of the Relation.

(Source: http://www.learndb.com/databases)

Let us demonstrate the normalization process by a certain case. We can dispense with redundancies and anomalies for the connection in Figure 143, in the event that we supplant it with the three relations in Figure 144, acquired by projections on the arrangements of ascribes comparing individually to the three items of data specified previously. The three relations are in Boyce–Codd normal form. Note that we have built three relations so each dependency compares to an alternate connection, the key of which is really the left hand side of a similar reliance. Along these lines, the fulfilment of the Boyce–Codd normal form is ensured, by the meaning of this normal form itself.

In the case we are analysing, the partition of the conditions (and along these lines of the properties actually represented by them) is encouraged by the structure of the dependency itself, "normally" isolated and independent of the others. In many cases, the deterioration can be completed by delivering the same number of relations as there are useful conditions (or rather, the useful conditions with various left hand sides). Tragically, a few conditions are observed to be perplexing. It may be pointless (or unthinkable) to construct the deterioration in light of the considerable number of conditions and it can be hard to distinguish the ones on which we should base the disintegration. We will illuminate this point by taking a gander at a basic case.

It is anything but difficult to confirm that the connection in Figure 145 fulfils the conditions 'Employee → Category' and 'Category → Salary'. By continuing in the way portrayed above, we could without much of a stretch acquire a database with 2 relations, both in Boyce – Codd typical shape. Then again, for a similar connection, we could have recognized both the practical conditions 'Category → Salary' also, 'Employee → Category Salary' (instead of 'Employee → Category'). Note that this dependency portrays the circumstance with pretty much a similar exactness (Ullman J. D., 1988).

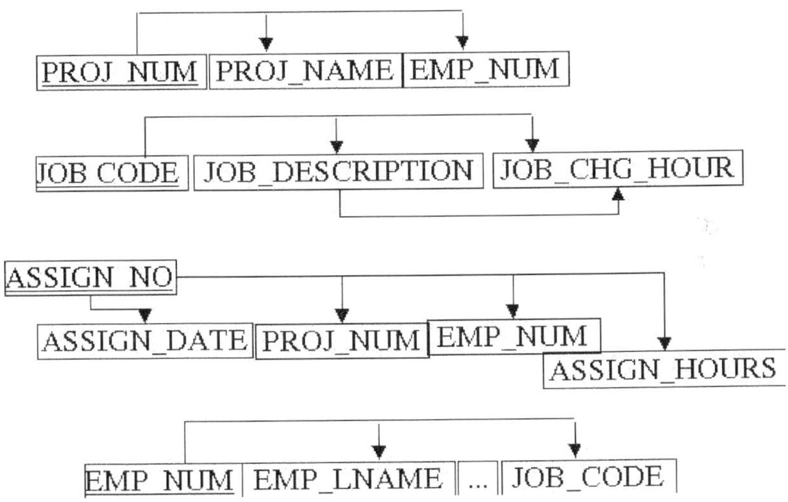

Figure 145 -

(Source: http://www.learndb.com/databases)

For this situation anyway, we would have had no indications on the most proficient method to create deterioration into Boyce–Codd normal form, in light of the fact that clearly the dependency 'Employee →

Category Salary' covers every one of the traits and in this way does not recommend any decomposed connection. This basic case demonstrates that the distinguishing proof of conditions can cause troubles with disintegration. We can envision what could happen when the connection has many traits also, a few useful conditions are characterized on it.

A total investigation of normalization would require, as an essential, a detailed research of the properties of the useful dependency. The finish advancement of these ideas goes past the extent of this book.

Given the significance of the subject, we expect to give a casual portrayal of the absolute most essential perspectives.

It ought to be noted, additionally, that we contemplate typical structures for the most part as assistant quality control instruments for relations and not as a plan method. The outline procedures talked about in going before sections are expected for the era of relations (in the legitimate outline stage) in view of entities and relationships that show up in the conceptual design composition (created amid the conceptual plan stage). The conceptual plan stage serves to recognize the principal ideas of the application to show, isolating them into unmistakable entities and relationships. Since the interpretation is then done considering every entity and every relationship independently, it is obvious that if the reasonable plan is done effectively, the relational model created amid the legitimate outline stage will be as of now normalized.

In this specific situation, normalization hypothesis is regardless helpful as a confirmation tool for the results of both applied and coherent plan.

DECOMPOSITION PROPERTIES

In this segment, we look at the idea of decomposition in more detail. We clarify how not all disintegrations are attractive and recognize some basic properties that must be fulfilled by a proper decomposition.

Lossless Decomposition

With a specific end goal to talk about the primary property, let us inspect the schema in Figure 146. This schema fulfils the functional dependencies:

$$\text{Employee} \rightarrow \text{Branch}$$

Project → Branch

Continuing similarly as some time recently, that is isolating on the premise of conditions, we will think that it is reasonable to decompose the schema into two parts:

- a schema Employee and Branch, relating to the dependency 'Employee → Branch';
- a schema Project and Branch, relating to the dependency 'Project → Branch'.

Let us look at the two relations in detail. Specifically, consider how it would be conceivable to reproduce data on the cooperation of the employees in the projects. The main probability is to utilize the Branch schema, which is the main ascribe normal to the two relations: we would thus be able to interface an employee to a project if the project is produced at the branch where the employee works. Lamentably, for this situation we are not ready to recreate precisely the data in the first relation: for instance the employee called 'Green works in Birmingham' and the Saturn project is being created in Birmingham, however in reality Green does not really work in the said project.

Functional dependencies:

Employee → Branch, Project → Branch

Figure 146 -

(Source: http://www.learndb.com/databases)

We can sum up the perception by taking note of that the decomposition of the unique relation must be helped out through a characteristic join of the two decompositions. Sadly, the natural join of the two decompositions Fig 146 produces the relation in Fig 146 that is not the same as the relation in Fig 145. The relation in Fig 146 contains all of the tuples of the first relation and additionally different tuples (the last two in the given table). The circumstance in the illustration relates to the general circumstance: given a relation r on a set of attributes X, if X-1 and X-2 are two subsets of' X' the union of which is equivalent to X itself, at that point the join of the two relations acquired by anticipating r on X-1 and X-2, separately, is a relation that contains all the tuples of r, in addition to conceivable others, which we can call ‘spurious’. Give us a chance to state that the decomposition of r on X-1 and X-2 is lossless if the actual join of the decompositions of r on X-1 and X-2 is equivalent to r-itself (not containing spurious-tuples).

Employee	**Project**	**Branch**
Brown	Mars	Chicago
Green	Jupiter	Birming-ham
Green	Venus	Birming-ham
Hoskins	Saturn	Birming-ham
Hoskins	Venus	Birming-ham
Green	Saturn	Birming-ham
Hoskins	Jupiter	Birming-ham

Figure 147 - Results for Figure 146.

(Source: http://www.learndb.com/databases)

It is obviously alluring, or rather an irreplaceable necessity that decomposition undertaken in order to achieve normalization is lossless.

We can recognize a condition that ensures the lossless decomposition of a relation as follows. Given that r is a relation on X and let X-1 and X-2 be subsets of X with the end goal that $X1 \cup X2 = X$. Besides, let $X0 = X1 \cap X2$. On the off chance that r fulfils the functional dependency $X0 \rightarrow X1$ or the functional dependency $X0 \rightarrow X2$, at that point the decomposition of r on X-1 and X-2 is lossless.

We can state that r has a lossless decomposition of two relations if there is a common attribute in the set of attributes of the two relations. In the case, we can see that the intersection of the attributes of the two relations is the Branch, which is not the left-hand side of any functional dependency.

We can legitimize the condition in the accompanying way, with reference to a relation r on the relation 'ABC' and to its decomposition A-B and A-C. Let us assume that r fulfils A→C. At that point, A is the key for the decomposition of the r on A-C furthermore, in this manner there are no two diverse tuples with similar values of A.

The join builds tuples in light of the actual tuples in the two decompositions. Let us consider a non-specific tuple t = (a, b, c) as a result of the join. We demonstrate that t has a place with r, so demonstrating the fairness of the two relations: t is gotten from t1 = (a,b) in the decomposition of r on A-B, while t2 = (a,c) in the decomposition of r on 'A-C'. As a result of decomposition, there must be two tuples in r, t1′ with values a and b on A-B, and t2′ with values a and c on A-C. Since r fulfils A→C, there is a solitary estimation of C in r related with the value on A: given that (a,c) shows up in the decomposition, this value is precisely c. In this way the estimation of t1 ′ on C will be c and in this way t1 ′ (which has a place with r) has values a, b and also c, and in this manner corresponds with t, which along these lines has a place with r, as we proposed to appear.

It is fitting to note quickly how the condition expressed is adequate however not entirely important to guarantee a lossless decomposition: there are relations that fulfil neither of the two conditions, but in the meantime they have a lossless decomposition. For instance, the relation in Fig 146 (acquired as the join in the decompositions) has actually a lossless decomposition on the two sets 'Employee Branch' and 'Project Branch'. Then again, the given condition guarantees that every one of the relations that fulfil a given arrangement of conditions have a lossless decomposition, and this is a valuable outcome: each time we decompose a relation into two relations that has common attribute, then the decomposition is lossless join decomposition.

Dependency Preservation

To present the second property, we can rethink the connection in Fig 145. Regardless if we wish to remove the anomalies, we could consider abusing just the dependency 'Employee → Branch', to acquire a lossless decomposition. We at that point have two relations, one on the properties Employee and Branch and the other on the properties Employee and

Project. The case in Fig 145 would be along these lines decomposed into the relations in Fig 148.

Employee	**Branch**
Brown	Chicago
Green	Birmingham
Hoskins	Birmingham

Employee	**Project**
Brown	Mars
Green	Jupiter
Green	Venus
Hoskins	Saturn
Hoskins	Venus

Figure 148 - Decomposition for Figure 145.

(Source: Teorey, T. J. Design of Database Structures)

The join of the two relations in Fig 148 produces the relation in Figure 149, for which we can state that the relation has a lossless decomposition on 'Employee Branch' and 'Employee Project'. This is affirmed by the way that Employee is the key for the relation. Tragically, the decomposition in Fig 148 produces another issue. Assume we wish to embed another tuple that determines the cooperation of the employee named Armstrong, who actually works in Birmingham, on the Mars project. In the unique connection, that is, the one in Fig 145, a refresh of this kind would be instantly distinguished as unlawful, in light of the fact that it would cause an infringement of the 'Project → Branch' dependency. On the decomposed relations anyway, it is definitely not conceivable to uncover any infringement of dependency. On the relation over Employee and Project, it is really impractical to characterize any functional dependency. Furthermore, in this manner there can be no infringement to appear, while the tuple with the qualities Armstrong and Birmingham fulfils the dependency 'Employee → Branch'. We can accordingly take note of how it is impractical to complete any check on the dependency 'Project → Branch', in light of the fact that the two characteristics Project and Branch have been isolated: one into one relation and one into the other one.

We can presume that, in every one of the decompositions, each of the functional dependency of the first mapping are ought to include qualities that seem all together in one of the decomposed patterns. Along these lines, it is conceivable to guarantee, on the decomposed pattern, the fulfilment of similar requirements as the first composition. We can state that a decomposition that fulfils this property safeguards the conditions of the first outline (Kimball, 1996).

Qualities of Decompositions

To outline the focuses talked about above, we can express that decompositions ought to dependably fulfil the properties of lossless decomposition and dependency preservation.

- Lossless decomposition guarantees that the data in the first connection can be precisely reproduced, remade without spurious data, in view of the data spoke to in the decomposed relations. For this situation, by querying the decomposed relations, we acquire similar outcomes that we would get by querying the original relation.
- Dependency preservation guarantees that the decomposed relations have the same ability to speak to the uprightness limitations as the original relation and in this way to uncover unlawful updates: each permitted update (individually, illicit) on the original relation compares to a permitted update (individually, illicit) on the decomposed relations. Clearly, we can have updates on the single decomposed relations, which have no partner in the original relation. Such updates are outlandish on non- normalized relations where they are wellsprings of anomalies.

Accordingly, from here on we will consider just decompisitions that fulfill these two properties. Given a decomposition that damages an ordinary frame, the standardization action is along these lines went for acquiring a decomposition that is lossless, jam the conditions and contains relations in normal form.

THIRD NORMAL FORM

Definition of Third Normal Form

The point of getting a proper decomposition into Boyce–Codd normal form can be accomplished. Infrequently, be that as it may, this is unrealistic, as we can see from a case. Take a gander at the connection in Fig 149.

We can expect that the accompanying conditions are characterized:

Manager	Project	Branch
Brown	Mars	Chicago
Green	Jupiter	Birmingham
Green	Mars	Birmingham
Hoskins	Saturn	Birmingham
Hoskins	Venus	Birmingham

Figure 149 - Decomposition with Problems.

(Source: Teorey, T. J. Design of Database Structures)

- 'Manager → Branch': every manager works at a specific branch;
- 'Project Branch → Manager': each project has more managers, who are in charge of it, however in various branches, and every manager can be in charge of more than one project; nonetheless, for each branch, a project has just a single manager in charge of it.

The connection is not in Boyce–Codd normal form in light of the fact that the left-hand side of the 'Manager → Branch' dependency is not a super-key. In the meantime, we can take note of how no great decomposition of this relation is conceivable; the dependency 'Project Branch → Manager' includes every one of the properties and in this way no decomposition can protect it. The case demonstrates to us that patterns exist that violates Boyce–Codd normal form and for which there is no decomposition that jam the conditions. We can in this way express 'Boyce–Codd normal from cannot be accomplished'. In such cases, we can however set up a less prohibitive condition, which permits circumstances for example, the one portrayed above, however does not permit additionally wellsprings of excess and inconsistency.

This condition characterizes another normal form: we will state that a connection r is in third normal from if, for each non-trivial functional dependency X → Y exists, at least one of the following is true:

- X contains the key K of the r (X is a superkey);
- Each one of the traits in Y is a member of some candidate key of r.

Coming back to our illustration, we can without much of a stretch confirm that, regardless of the possibility that the mapping does not fulfil the Boyce–Codd normal form, it fulfils the third normal form. The 'Project Branch → Manager' reliance has as its left-hand side the key for the connection, while 'Manager → Branch' has an exceptional quality for the privilege hand side, which is a piece of the 'Project Branch' key. The relations demonstrate a type of excess: each time a manager shows up in the tuple, the branch for which he or she works is rehashed. This excess is "endured" by the third normal form, in light of the fact that naturally, a decomposition that dispensed with such excess and in the meantime protected all the conditions would not be conceivable.

The third normal form is less prohibitive than the Boyce–Codd normal form , hence does not offer similar assurances of value for a relation; it has the favorable position be that as it may, of continually being achievable.

The name of this normal form proposes the presence of other normal forms, which we will manage quickly. The main normal form just builds up a condition that is at the premise of the relational model itself: the characteristics of a connection are characterized on nuclear esteems and not on complex values, regardless of whether sets or relations. The second normal from is a frail variety of the third, which permits useful, conditions, for example, that among Category and Salary of the connection. Note that this dependency fulfills neither of the states of the third normal form. There are without a doubt other typical structures that allude to other honesty limitations.

None of these typical structures is utilized essentially in current applications since the third normal form and the Boyce–Codd normal form as of now give the right bargain among straightforwardness and the nature of results.

Decomposition into Third Normal Form

Decomposition into third normal form can continue as recommended for the Boyce–Codd normal form. A connection that does not fulfil the third normal form is decomposed into relations acquired by projections on the properties comparing to the functional dependencies. The main condition to ensure in this procedure is of continually keeping up a relation that contains a key to the original relation. We can see this by alluding to the relation in Fig 150, for which we have a full functional dependency, 'Employee → Salary'.

Employee	**Project**	**Salary**
Brown	Mars	50
Green	Jupiter	30
Green	Mars	30
Hoskins	Saturn	40
Hoskins	Venus	40

Figure 150 - Relation for the Third Form.

(Source: Teorey, T. J. Design of Database Structures)

Decomposition in the relation on the attributes 'Employee Salary' and in another on the sole attribute Project would abuse the property of lossless decomposition, basically in light of the fact that neither of the two relations contains a key for the first connection. To ensure this property, we should characterize the second relation on the attributes 'Employee Project', which really frames a key for the first connection. We demand however on the way that the achievement of the decomposition depends generally on the conditions we have recognized.

To finish up, we come back to the case in Figure 150 and take note of that the relation does not fulfil even the third normal form. Continuing as proposed, we can in any case get the decomposition appeared in Figure 144, which, we saw, is likewise in Boyce–Codd normal form. This is an outcome that is legitimate in an extensive variety of cases: a decomposition with the aim of acquiring the third normal form frequently delivers outlines in Boyce-Codd normal form. In specific, we can demonstrate that if a relation has just a single key (as for this situation) at that point the two normal forms match. A relation with just a single key is in third normal form if and just on the off chance that it is in Boyce–Codd normal form.

Other Normalization Forms

Alluding to Figure 149, we will take a gander at promote thoughts regarding normal forms. By looking at the details all the more nearly, we see that we could have portrayed this piece of the application in a more reasonable way by presenting another attribute that is Division, which isolates the single branches as indicated by their manager. The relation is presented in Fig 151.

Manager	Project	Branch	Division
Brown	Mars	Chicago	1
Green	Jupiter	Birmingham	1
Green	Mars	Birmingham	1
Hoskins	Saturn	Birmingham	2
Hoskins	Venus	Birmingham	2

Figure 151 - Restructuring for Figure 149.

(Source: Teorey, T. J. Design of Database Structures)

The conditions can be characterized as takes after:

- 'Manager → Branch Division': every manager works at one branch and oversees one division;
- 'Branch Division → Manager': for each one the branches and divisions there is a solitary manager;
- 'Project Branch → Division': for each branch, a project is dispensed to a solitary division and has a sole manager; the functional dependency 'Project Branch → Manager' can be recreated.

For this mapping, there is a proper decomposition, as appeared in Fig 152:

- the decomposition is lossless, on the grounds that the regular attributes Branch and Division form a key for the first relation;
- the conditions are saved, in light of the fact that for every dependency there is a decomposition connection that contains every one of the attributes included;
- both the relations are in Boyce–Codd normal form, in light of the fact that for all the conditions the left hand side is comprised of a key.

We can in this manner finish up by expressing that, regularly, the trouble of accomplishing Boyce–Codd normal form could be expected to an inadequately exact investigation of the application.

Manager	Branch	Division
Brown	Chicago	1
Green	Birmingham	1
Hoskins	Birmingham	2

Project	Branch	Division
Mars	Chicago	1
Jupiter	Birmingham	1
Mars	Birmingham	1
Saturn	Birmingham	2
Venus	Birmingham	2

Figure 152 - Good Decomposition of Figure 151.

(Source: Teorey, T. J. Design of Database Structures)

DATABASE OUTLINE AND NORMALIZATION

The hypothesis of normalization, regardless of the possibility that concentrated in streamlined frame, can be utilized as a reason for quality control operations on patterns, in both the applied also, consistent plan stages. We will quickly remark on its utilization in consistent plan, and afterward represent the adjustment of the standards to the Entity-Relationship model and consequently to logical design.

An outline is now and again fragmented, thus a correction of the relations acquired amid the logical design stage can recognize places where the conceptual design can be refined. The confirmation of the plan is regularly moderately simple. This is on account of the distinguishing proof of the useful conditions and the keys must be done inside the setting of a solitary connection, which is gotten from an element or a relationship effectively broken down in the physical design stage. In this specific situation, the structure of conditions is for the most part rather straightforward and it is along these lines conceivable to distinguish specifically the decomposition important to acquire a normal form. For instance, the relation in Figure 151 would be created just if, amid the reasonable configuration stage, we do not understand that class and representative are free ideas. By distinguishing the presence of useful conditions, it is consequently conceivable to cure mistakes.

Verification of Normalization on Entities

The thoughts on which normalization is based can likewise be utilized amid the logical design stage for the quality control of every component of the logical design. It is conceivable to consider every entity or relationship as though it were a relation. Specifically, the relation that compares to an entity has credits that compare precisely to the attributes of the entity. For an entity with an outside identifier, additionally ascribes are important to incorporate the entities that take an interest in the distinguishing proof. It is adequate to consider the practical conditions that exist among the attributes of the entity and to check that each of them has the actual identifier as left-hand side. For instance, let us consider a PRODUCT entity, with Code, ProductName, Supplier, Address, and SupplierCode attributes. Supplier is the name of the supplier of a product, for which the Address and the SupplierCode are imperative.

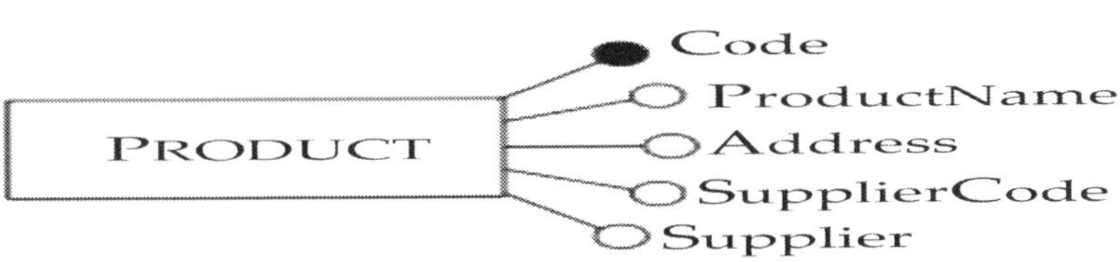

Figure 153 - Entity which Partakes in Verification.

(Source: Teorey, T. J. Design of Database Structures)

In distinguishing the conditions for this entity, we can take note of that different suppliers can exist with a similar product or address, while all the attributes of every supplier are distinguished by its SupplierCode. Along these lines, the dependency 'SupplierCode → Supplier Address' exists. Moreover, all the attributes are practically subject to the Code attribute, which effectively constitutes the identifier of the entity. Once a code has been settled, the product and the supplier are unambiguously decided, with their attributes. Since the main identifier of the entity is comprised of the Code only, we can presume that the entity violates the third normal form. This is on the grounds that the dependency 'SupplierCode → Supplier Address' has a left hand side that does not contain the identifier and a right-hand side made up of attributes that are not some portion of the key. In these cases, the trial of normalization demonstrates that the reasonable composition is not precise and recommends the decomposition of the entity itself.

Verification of Normalization on Relationships

In concerning relationships, the thinking is much less complex. The arrangement of events of every relationship is a connection, and along these lines it is conceivable to apply the normalization systems straightforwardly. In any case, the areas on which the connection is characterized are the arrangements of events of the entity included.

Subsequently, to confirm that a normal form is fulfilled, we should recognize the existing functional dependencies among the entities included. Since it is simple to demonstrate that every parallel connection is in third normal form (and furthermore in Boyce–Codd normal form), the check of normalization is done just on the n-ary relations, on those which include no less than three substances. Consider for instance, the relationship THESIS in Figure 154, which includes the entities STUDENT, PROFESSOR, DEGREEPROGRAMME, and DEPARTMENT.

It depicts the way that students, selected on degree programs, compose thesis in department under the supervision of professors.

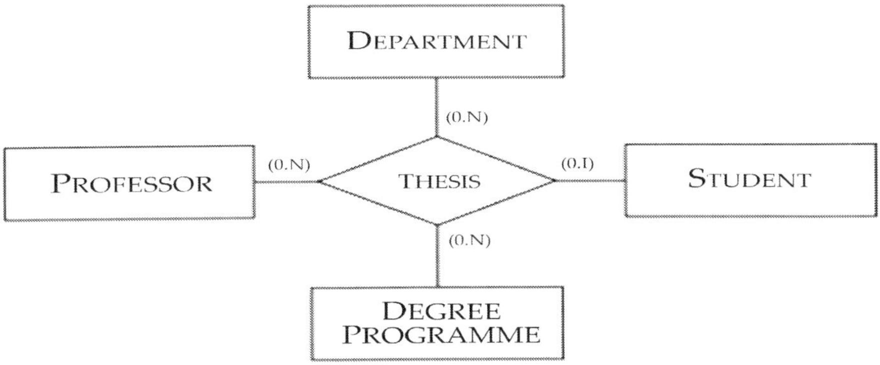

Figure 154 - Relationship with Normalization which is yet to be Verified.

(Source: Tsichritzis, D. L. Data Models.)

We can touch base at the accompanying conclusions:

- every student is selected on a degree program;
- every student composes a thesis under the supervision of a professor (who is not really identified with the degree program);
- every professor has a place with a solitary department and the students under his/her supervision compose their thesis under the care of that department.

Give us a chance to assume that for the reasons for the postulation, the professor's office is not significant to the degree program on which the student is enlisted. We would then be able to state that the properties of the application are totally depicted by the accompanying three functional dependencies:

$$STUDENT \rightarrow DEGREEPROGRAMME$$

$$STUDENT \rightarrow PROFESSOR$$

$$PROFESSOR \rightarrow DEPARTMENT$$

The super key of the connection is STUDENT: given a student, the degree program, the professor, and the department are unambiguously recognized.

Therefore, the third functional dependency causes an infringement of the third normal form. The association of a professor to a department is an idea free of the presence of students who compose thesis with the professor. Thinking as some time recently, we can reason that the relationship presents undesirable viewpoints and that it ought to be decomposed, isolating the useful conditions with various left hand sides. Along these lines, we can acquire the mapping in Fig 155, which contains two connections, both in third normal form (and in Boyce–Codd normal form). Here likewise, we have a lossless decomposition with safeguarding of conditions.

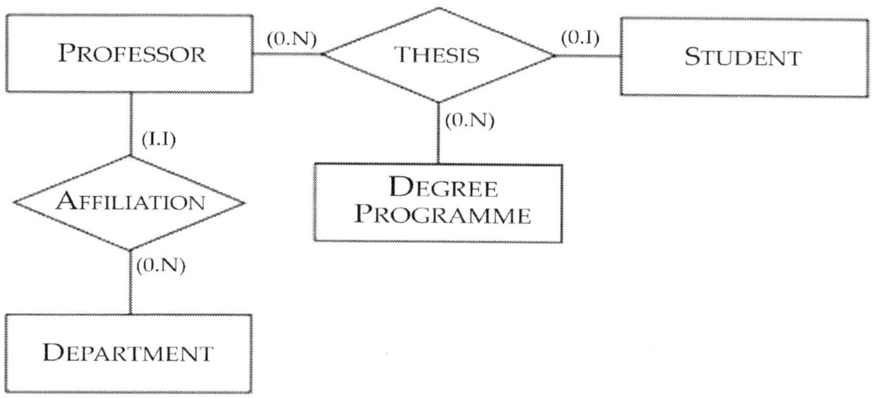

Figure 155 - The Results for Decomposition.

(Source: Tsichritzis, D. L. Data Models.)

Further Decomposition of Relations

On the outline in Fig 155, we can mention some further objective facts. These perceptions go past the hypothesis of normalization in its strictest sense, however stay inside the field of investigation and confirmation of theoretical patterns by methods for formal devices, in the particular instance of useful conditions. The relationship THESIS is in third normal form, since its key is comprised of the STUDENT entity, and the most important conditions that actually exist on it are those that have this entity as left hand side, in particular, 'STUDENT → PROFESSOR' and 'STUDENT → DEGREEPROGRAMME'. Then again, the attributes portrayed by the two conditions are independent of each other. Not all students are composing thesis thus not every one of them have professor. From the standardization point of view, this circumstance does not present issues, since we accept that the relations can contain invalid values, given that they are not in the key. Subsequently, it is sensible to acknowledge conditions with a similar left hand sides.

At the theoretical demonstrating level, we should recognize among the different ideas. In addition, there is no understanding of 'invalid values in a relationship', nor would there actually be any real sense in presenting one. Utilizing the conditions, we can consequently reason that it is suitable to break down the relationship further, getting two connections, one for each of the two ideas. Fig 156 demonstrates the decomposed design. The decomposition is worthy for this situation additionally, preserves the dependency and is lossless. On the off chance that we sum up the contention created above, we infer that it is fitting to decompose the n-ary relations on which there is a dependency whose correct hand side contains more than one entity. Since it is uncommon to experience connections that include more than three elements, we can say that it is typically helpful to decompose any ternary relationship on the off chance that it has a functional dependency whose left-hand side comprises of one entity and the right-hand side comprises of the other two.

For illustration, if the two entities in the right-hand side of the dependency are firmly connected to each other or if there really exist different conditions that would not be saved in the decomposition, for example, in the event that we are intrigued just in students who are as of now composing their thesis. Hence, for each one of them, we have a degree program and a professor who is administering his or her thesis.

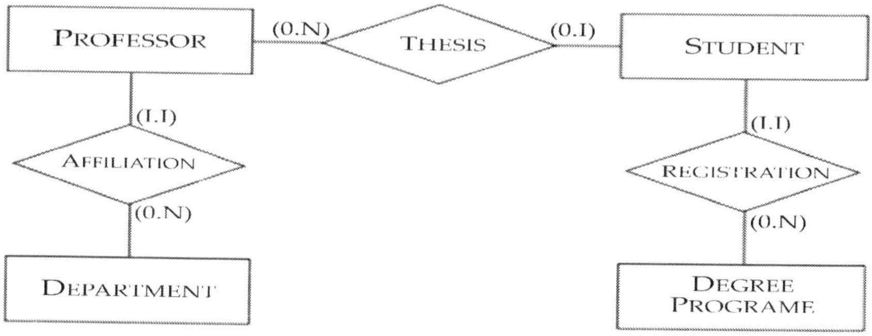

Figure 156 - Further Decomposition Results.

(Source: Tsichritzis, D. L. Data Models.)

CHAPTER 10

APPLICATIONS - DATABASES AND THE WORLD WIDE WEB

The advancement of the Internet is a standout amongst the most critical wonders in data innovation. The clients are quickly expanding in number and assortment, from organizations with fast systems to people with moderate modem associations. Organizations frequently orchestrate private systems that utilize similar conventions and tools created and utilized on the Internet. These are called intranets and are associated with the Internet. A noteworthy innovation on both Internet and intranets is the World Wide Web (WWW), regularly called just the Web. Tthe Web has presented another worldview for both the dispersal and obtaining of data, which is anything but difficult to utilize, adaptable, and efficient. The Web was initially an interface for access to dispersed records. It is currently a stage for data frameworks of various kinds, having in like manner just the reality of being available, specifically or in a roundabout way, through nonexclusive instruments, called programs. Frameworks of this sort are frequently called Web Information Systems (WIS); coupled to DBMSs, they can bolster and give access to a lot of information. Thus, the incorporation of Web-based innovation for information administration is basic. In the meantime, WISs likewise show particular issues. In any case, the sort of upheld data is profoundly heterogeneous. Since the beginning, the Web has been utilized to oversee printed and media data (pictures, sound, video). Conversely, customary data frameworks for the most part oversee information that can be spoken to by methods for basic information structures: sets of tuples with nuclear esteems. In outline, the structure of the data available in the Web is more intricate and less organized and customary than that utilized as a part of databases. Another recognizing highlight of WISs is that they regularly offer more noteworthy adaptability to the client in the connection with the systems: cooperation is driven by the enthusiasm of the client instead of following pre-composed examples. Conversely, clients have constrained open door for changing the information content. For example, in electronic trade, clients can control just the substance of their private own 'basket', containing the deal points of interest of the obtained merchandise but cannot adjust the substance of the Web information depicting offers. Moreover, the connection between a Web program and a wellspring of data makes utilization of instruments that are not quite the same as those utilized as a part of customary system. In spite of the fact that the considerable institutionalization of programs makes the interfaces genuinely uniform, their qualities can differ, contingent upon the gadget being utilized for getting to the Web. Think about the distinctions that can exist between the abilities accessible on a capable workstation with an extra-large screen on a fast intranet and that of a convenient PC with a little screen and a cellphone association. Also, applications are offered to a substantial group, all the Internet clients. In this way, the objectives to be considered are significantly more extensive and more different than in conventional data frameworks. For every one of these reasons, a re-examining of the WIS improvement process itself is required. This will apply both to the parts expressly created and to those to be coordinated into them. In this section, we will take a look at the collaboration between the methods of databases and those of the Web. In the wake of reviewing the basic ideas of the Internet and the Web, we will outline the development of WIS applications and the fundamental classifications into which they can be arranged. In the two after areas we will manage the viewpoints that are most pertinent to databases. Initially, we address the issues concerning the display and outline of Web locales, by concentrating on those Web destinations in which information assumes an imperative part. We call them information serious Web destinations and we trust that their significance will increment essentially sooner than later. At last, we will manage the innovation for incorporating Web destinations and databases.

THE INTERNET AND THE WORLD WIDE WEB

We will give a short review of those parts of the Internet and the World Wide Web that are essential for whatever remains of the section.

The Internet

Internet can be characterized as an alliance of systems that convey by methods for a similar arrangement of conventions, those of the TCP/IP (Transmission Control Protocol/Internet Protocol) family. As a rule, a hub of the Internet is a piece of a neighbourhood arrange, which interconnects workstations or PCs situated inside a little range. Nearby systems speak with each other inside system pecking orders, for instance the system of a college, the system of the considerable number of colleges in a nation, a general national system, et cetera.

Every hub on the Internet has an IP address, which exceptionally recognizes it. IP is composed of four numbers, for instance 131.175.21.1. For the most part, a machine on the Internet additionally has a representative name, made up of identifiers isolated by specks, which can be utilized instead of the IP address. For instance, the emblematic name 'morgana.elet.polimi.it' can be utilized to allude to the IP address. From the address structure, the government arrange association ends up plainly straightforward, as the client (individual or program) can allude to any hub with a known address. From the intelligent perspective, every Internet hub can associate straightforwardly with some other hub. Correspondence happens by methods for a layered association, which includes different conventions, including the TCP/IP, in which points of interest is beyond the scope of this book. A basic normal for every one of the applications that work on the Internet (and for the most part on whatever other system that uses the TCP/IP conventions) is the selection of the customer server worldview. The customers deal with the connection with the client and the servers do the asked for operations, furnishing the customers with fitting reactions. Servers offer an arrangement of predefined capacities (specifically, Web administrations), in light of standard conventions, which utilize the administrations given by TCP/IP.

The World Wide Web

Give us a chance to present the World Wide Web through a progression of steps. A hypertext is an archive with a non-successive structure, made up of different parts, related by methods for joins. The segments can be straightforwardly gotten to by following the connections, without the unbending nature of the consecutive physical structure. To give a first case, let us represent how the substance of a book could be sorted out by methods for a hypertext, with a various levelled structure. The root contains a general depiction (for instance the prelude) and enables access to succeeding parts, each of which is as yet sorted out with a conceivable brief acquaintance and connections with its areas. Over the span of the content, there can be references to some point (as in a conventional book there are references to pages or segments) or to a list of sources. The idea of hypertext can be summed up in different ways. In any case, we have talked about a solitary archive, while a similar method can be utilized to relate many records, started by various individuals at various circumstances. From the reasonable perspective, there is no distinction between the inward connections of a record and those to different archives. Moreover, the reports can be literary as well as of different sorts, for example, pictures, video or sound. For this situation, we allude to an interactive media hypertext, regularly condensed to hypermedia. At long last, as we are alluding to the Internet, the arrangement of records that make up the hypermedia can be conveyed over numerous hubs (see the outline in Figure 157). To compress, we can state that the World Wide Web is a circulated sight and sound hypertext with autonomous parts, as it interfaces records of different sorts created and kept by various subjects over the Internet.

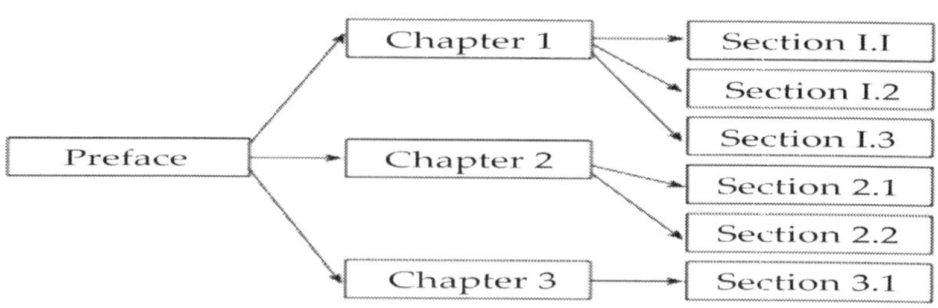

Figure 157 - Hierarchical Structure Hypertext.

(Source: Ceri, S. Methodology and Tools for Database Design.)

Not only that the Web enable access to static records as portrayed above, but also offers the chance to dispatch programs that powerfully produce new pages. In this way, the era of a given page can be founded on substance removed from a database. The major mechanical segments of the Web, which permit the usage of the thoughts that we have portrayed above, are shown underneath. They are the html (Hypertext Markup Language), the URL (Uniform Resource Locator) idea for referencing assets, and the HTTP (Hypertext Transfer Protocol) correspondence convention.

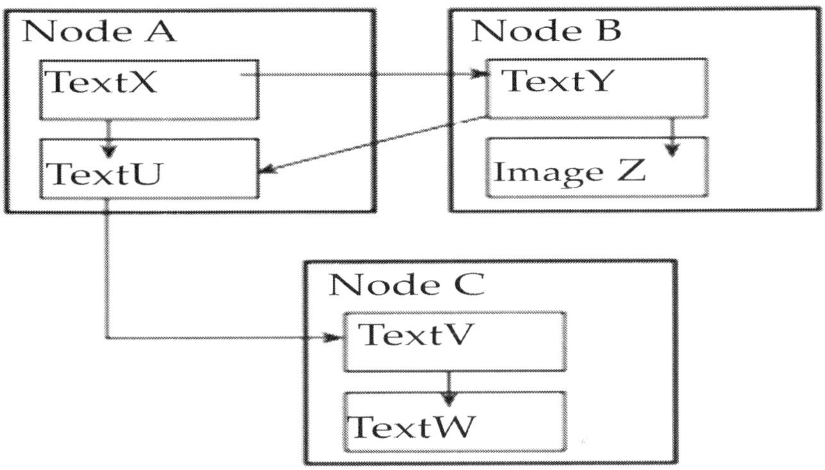

Figure 158 - Multimedia Hypertext.

(Source: Ceri, S. Methodology and Tools for Database Design.)

INFORMATION SYSTEMS ON THE WEB

We said toward the start of this section the Web is a general interface for access to data 1frameworks of different kinds. We foresee that sooner than later it will end up being the standard interface, or possibly the common one. Unquestionably, the quantity of information systems on the Web and their capacity will increment quickly. Keeping in mind the end goal to get ready for a talk on the part of databases in WISs, we have to distinguish the primary classes of WIS. Before going into the discourse, a perception is fitting. The Web began in the logical world, and it spread first through individual clients and not proficient ones. Just later it turned into a working instrument. A term regularly used to portray utilization of the Web is surfing, which proposes a use that is shallow and recreational. We will concentrate on the Web as a gadget for seeking after processional or business situated targets with regards to databases, where surfing is not a focal issue. Nonetheless, information serious applications offer expansive accumulations of information to end clients, accordingly supporting types of surfing.

Publishing on the Web

The first goal of the Web was to make information accessible. Numerous associations utilize the Web for making information accessible to all clients. This information is composed in HTML pages that are purposively constructed, albeit a large portion of them depend on prior data. The clients of these frameworks counsel the data distributed on the Web. As the request and the amount of distributed data builds, it ends up noticeably clear that the manual administration of HTML records is troublesome.

- Data may change. These progressions should be engendered to the information substance of the Web page. Information is interconnected with the hypertext association and the realistic introduction; this makes Web site advancement more troublesome.
- The Web page can store information in an excess form, with a specific end goal to encourage neatness and route. For instance, consider the pages about college educators. On each of these pages there could be not just a connection to the fundamental page of the division, yet additionally the name and address of the educator, which is the same as the address of the office.

On the off chance that the division changes its address, the alteration would in this way influence many pages. These two issues are addressed if data are stored in a database. This enables the pages to be built automatically, bolsters the partition of the different perspectives and wipes out the issues of repetition. The second primary explanation behind the connection of Web destinations and databases is simply the way that the data may be stored databases. In these cases, the purpose behind distribution on the Web is to empower simpler access to a lot of pertinent data, which can be distributed on the Web at generally little cost.

Transactions on the Web

At first, the databases open through the Web were refreshed by different means, for example, effectively existing applications or particular capacities for stacking and changing information, oversaw locally. A characteristic stride now was to take into account refreshes through the Web, conceivably through pages fundamentally like those utilized for show, so disseminating the work to various people. For instance, in a college site, the updates concerning the staff of every division could be overseen by the organization of the office itself. Access by methods for the program hence substitutes for other UIs, bringing the advantage of consistency.

Electronic Business and New Applications

Aside from the expanded number of clients and the consistency of the interfaces, the applications depicted above can be thought of as customary with regards to data frameworks. Then again, the Web has created numerous new applications, which are conceivable because of the soul of joint effort and investment that energizes a large portion of the Internet clients. A standout among the most imperative and testing new applications is electronic business, the utilization of the Web for offering merchandise and ventures. Many organizations offer items on the Web, for instance, books, programming instruments, air tickets, and so on. Data about offered items must be sorted out and stored at the organization's webpage, regularly by methods for the DBMS. This application is to a great extent an augmentation of existing systems: customary mail arranges is frequently in light of data frameworks and databases, which are utilized by organization. Presently, with Web-based frameworks, get to is offered specifically to the buyer. Plainly, the security issue ends up noticeably basic now, particularly for the instalment for deals, which is typically done by conveying credit card numbers on the Web. A few other new uses of Web innovation are less firmly associated with database innovation, in spite of the fact that they may require the help of a DBMS for the administration of vast collection of informational. The exchange gather thought has been taken up by business associations. For instance, the organizations that offer books through the Web typically offer the likelihood of composing remarks and amendments, which wind up noticeably open property. The administration of the data can be done by methods for basic documents, at the same time, if the size expands, it can be helpful to have the help of a database. Facilitate improvements of the potential offered by the Web are influencing different sorts of frameworks. For instance, bolster is being created for types of co-agent work. The thought has just been across the board in the specialized and academic group for a considerable length of time, and concerns tools, specifically programming, ready to bridle the energy of individual or gathering movement, otherwise called Computer-Supported Cooperative Work. The most progressive types of co-agent frameworks require the controlled sharing of archives and the co-appointment of exercises and work process. The product devoted to this reason, known as work process administration frameworks, empowers the association of work inside workplaces by controlling exercises and the stream of the records that they deliver. The Internet gives an exceptionally valuable foundation to the formation of work process frameworks and the Web is the common interface. It ought to be underlined, in any case, that a work process frequently requires a level of organizing exercises and support of incorporated and particular programming frameworks that can be in struggle with the open idea of the Web. We can state that these applications are normally offered to a specific scope of clients who co-work as indicated by particular principles and make utilization of particular devices that are not accessible. Consequently, work processes are frequently utilized over intranets. Notwithstanding, a similar framework ought to be utilized for both secured and open applications. This need prompted the improvement of extranets, systems that are controlled by given associations, but at the same time are interested in some type of outer get to. On extranets, clients are characterized and every client is empowered to utilize distinctive applications. An established illustration is the extranet offered by expedited service organizations, where data is given by the organization's representatives, however clients can screen the conveyance of their mail by utilizing a similar Web interfaces. Another unique part of the Web is the likelihood of offering coordinated administrations, took care of by various subjects. Co-operation here can be free or unbending. A case of the previous is the 'rundown of valuable connections' or 'related destinations', which are found on many pages. The second case is actually an interface that permits bibliographical hunt over many locales. A last case is the action of information cultivating: given the expansive measure of information accessible on the Web, an organization or element can get and sort out data through the Web, perhaps for reconciliation into its information distribution centre. An alternate heading that is creating in WISs is that of inserted frameworks. These frameworks do not utilize conventional terminals or PCs as interfaces, but utilize a range of gadgets, such as cellphones, remote palmtop PCs, information accumulation gadgets, or control frameworks for mechanical establishments. A typical for these frameworks is the constraint of the assets (registering power, memory, data transfer capacity, and size of display). As of now, this kind of framework presents couple of parts of particular enthusiasm from the perspective of databases. Then again, soon we expect a quick development of "terminals" associated with the Internet, joined into vehicles or household machines for instance, and afterward the interest for database communication will increment. For instance, consider a framework in a vehicle that uses a GPS gadget to decide the position of the vehicle on any surface of the Earth surface and acquires from the Web climate or range of visitor data. Then again, for residential tools, think about a framework for helping out upkeep and programming refreshing through the Internet (Graham, 1996).

DATA-INTENSIVE WEBSITES

In this area we will quickly talk about how the methods for the plan of databases can be adjusted for the outline of Web destinations. Obviously, the outline of Web destinations is an unpredictable action and cannot be managed in a couple of pages. In this way, we will give a review of the fundamental issues. Give us a chance to clear up the degree of the outline movement that we are presently going to consider. One could think about the whole Web as a brought together wellspring of data. This is not a supportive view, as there is no cognizance in the frame in which the data is offered, in its quality, substance or ease of use. At the other extraordinary, we could consider the Web as an arrangement of 'free pages', each of which constitutes a particular wellspring of data. This is the perspective of the arrangements of bookmarks oversaw by programs and of web indexes, which select connections to single pages. The two options are clearly wrong to set the extent of Web outline. Indeed, outline needs to focus on a framework that is devoted to a particular application requirement and under the control of a solitary subject, or if nothing else of different facilitated subjects who share a similar plan objectives. Since Web destinations ordinarily appreciate this property, we set Web locales as the degree of the outline action. We additionally focus our thoughts regarding information serious Web locales. That is, destinations whose basic role is to convey a lot of data to an assortment of clients. These requirements are best served by locales with extremely customary structures. For instance, the different divisions of a college, each with its own educators and courses, regularly have comparable Web pages, homogeneously interconnected. Correspondingly, a database may store data about educators and courses. Given the normality of Web pages, it ends up noticeably helpful to configuration bits of the Web with procedures like those used to plan databases. In this manner, our consideration will be fundamentally focused on destinations with no less than a specific level of normality in the data offered, whose objective can be depicted as the 'distribution of the substance of a database' for conference purposes (Greenspun, 1999).

Give us a chance to consider an organization that has a database with the applied pattern and the coherent diagram in Figure 159. (This is an improved adaptation of a case demonstrated before.) Suppose that the organization wishes to sort out a Website that offers the substance of the database for discussion. A streamlined, yet sensible association for this site incorporates:

- a home page, with connections to all the branches and two different pages, containing connections to every one of the projects and to every one of the employees of the organization, individually;
- a page for each branch, with a rundown of departments, each with a connection to the individual administrator and a connection to another page with a rundown of employess of a department;
- a page for each project, with a rundown of connections to the employees in the project;
- a page for every employment, with individual data and connections to the branch and to the projects in which he or she is included.

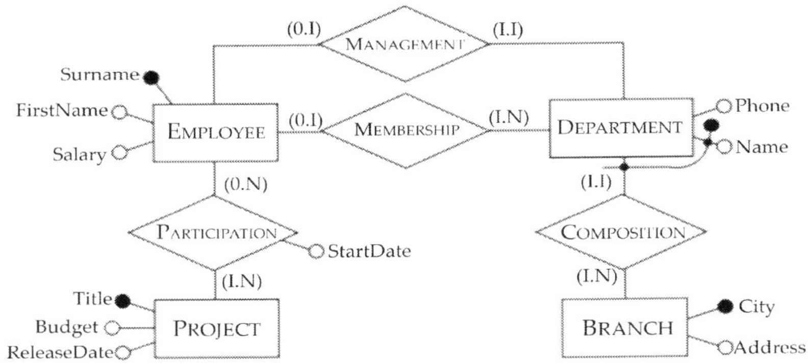

EMPLOYEE(*Surname*, Firstname, Branch, Department)

DEPARTMENT(*Branch*, *Name*, Phone, Manager)

BRANCH(*City*, Address)

PROJECT(*Title*, Budget, ReleaseDate)

PARTICIPATION(*Employee*, *Project*, StartDate)

Figure 159 - E-R Schema and Logical Schema of the Database.

(Source: Ceri, S. Methodology and Tools for Database Design.)

TECHNIQUES FOR DATABASE ACCESS THROUGH THE WEB

In this area, we will display an outline of the procedures that can be utilized for interfacing a Website and a database. We will start by displaying the most widely recognized approach, in light of the CGI convention.

Database Access through CGI programs

```
<html>

<head><title>John Smith</title></head>

<body>
```

```
<H3>John Smith</H3>
<table>
<tr><td><em>Department</em> :</td><td> Design</td></tr>
<tr><td><em>City</em> :</td><td> London</td></tr>
<tr><td><em>Address</em> :</td><td> 354 Regent St</td></tr>
<tr><td><em>Salary</em> :</td><td> 45,000 per year</td></tr>
<tr><td><a href="/cgi-bin/Projects?Surname=Smith">Current
        projects</a></td><td></td></tr>
</table>
</body>
</html>
```

Figure 160 – HTML Code.

(Source: http://www.webbasedprogramming.com/)

The most straightforward method for access to a database from inside the Web comprises of the CGI convention. The strategy is general and the rule on which it is based is basic. The HTTP server gets a demand, perceives that the asset showed by the URL's program and dispatches this program through the CGI convention. The program gets to the database and returns the outcomes to the server, which at that point sends them to the program. We have here two separate situations, the HTTP server and the application program. They collaborate through an interface, which is exceptionally straightforward and all around characterized. Expect now that we are keen on producing, by methods for a CGI program, the pages for the employees as appeared in Figure 161.

```
http://www.nc.com/cgi-bin/Employee?Surname=Smith
```

```
main(char Surname[])
{
  char Firstname[20], Department[20], City[20];
  char Address[60];
  int Age, Salary;
  $ open connection to NewCompanyDB
  $ select Firstname, Department, City, Address, Salary
      into :Firstname, :Department, :City, :Address, :Salary
    from Employee E, Branch B
    where E.Branch = B.City
      and Surname = :Surname ;
  $ close connection
  if (sqlcode == 0){
    printf("<html>\n<head><title> %s %s",Firstname,Surname,
           "</title></head>\n<body>\n");
    printf("<H3> %s %s",Firstname,Surname,"</H3>\n");
    printf("<table>\n");
    printf("<tr><td><em>Department</em>:</td><td>%s",
           Department,"</td></tr>\n");
    printf("<tr><td><em>City</em>:</td><td>%s",City,
           "</td></tr>\n");
    printf("<tr><td><em>Address</em>:</td><td>%s",Address,
           "</td></tr>\n");
    printf("<tr><td><em>Salary</em>:</td><td>%u",Salary,
           "</td><tr>\n");
    printf("<tr><td><a href=\"/cgi-bin/ProjOfEmp?Surname=%s",
            Surname,"\">Current projects</a></td>
            <td> </td></tr>\n");
    printf("</table>\n</body>\n</html>");
    }
  else {
    printf("<html>\n<head><title>Not found</title></head>
           \n<body>\n");
    printf("No employee has surname %s\n",Surname,"</body>\n
           </html>");
  }
}
```

Figure 14.12 A CGI program that produces pages as in Figure 14.8.

Figure 161 - A CGI Program.

(Source: http://www.webbasedprogramming.com/)

We demonstrated a straightforward CGI program written in C, with inserted SQL, that takes a surname as information and creates the page of the employee with that exact surname. The program can be rung utilizing a URL that incorporates the surname of the employee as info parameter.

Since the Surname is the real key of the connection, it is ensured that the select statement returns at most one tuple thus it is not important to utilize cursors. The program opens the association with the database, executes the question, stores the outcome in reasonable nearby factors and creates the page. This page is sent to the http server, which at that point sends it to the customer. The HTML design is indicated by methods for the constants in the yield explanations.

Development Tools

We quickly said in Chapter 1 that the presence of tools helps database advancement, specifically as respects the generation of segments of UIs, such as structures, menus, and reports. The necessities of Web applications have caused the fast appearance of committed advancement tools that have similar objectives. The principal thought is to encourage the development of Web pages in view of the substance of the databases. HTML pages can be created by these instruments in view of a prefixed association. Numerous items offer the office of utilizing HTML formats, which are the skeletons of the HTML pages with a segment of the substance, characterized utilizing an SQL inquiry, potentially with parameters.

```
<html>
<! tmplSQL connect database NewCompanyDB>
<! tmplSQL query "select Firstname, Department,
                         City, Address, Salary
                  from Employee E, Branch B
                  where E.Branch = B.City
                  and E.Surname = $Surname
                 ">
<head>
<title></title>
</head>
<H3>$1 $Surname</H3>
<table>
<tr><td><em>Department</em> :</td><td> $2</td></tr>
<tr><td><em>City</em> :</td><td>$3</td></tr>
<tr><td><em>Address</em> :</td><td>$4</td></tr>
<tr><td><em>Salary</em> :</td><td> $5 per year</td></tr>
<tr><td><a href="/cgi-bin/tmplSQL?tmpl=ProjOfEmp&Surname=$Surname">
        Current projects</a></td><td> </td></tr>
</table>
</body>
</html>
```

Figure 14.13 An HTML template for the construction of pages such as those in Figure 14.8.

Figure 162 - HTML Template.

(Source: http://www.webbasedprogramming.com/)

The format creates a similar html source as the CGI program with the one showed before, yet without utilizing a program in C. The HTML code is not covered up in the yield proclamations, but rather specifically determined. This is helpful while investigating the structure of pages: a program disregards the HTML components of the frame and would thus be able to demonstrate the layout, with the names of the factors ($1, $2, et cetera) set up of the individual esteems. The tools that permit the utilization of layouts work in a way that is like CGI programs: the URLs mean a particular program, which we could call the 'format supervisor', that gets as info the name of the layout and the required parameters (in the case, the surname), gets to the database and returns an HMTL page as the outcome. The format in the figures above can be summoned with a URL.

Numerous other improvement tools have been proposed, predominantly to help access to Web destinations utilizing devices like those accessible for customary database applications.

REFERENCES

1. World Economic Forum. (2010). *The Global Information Technology Report 2009–2010.* Geneva.
2. Abiteboul, S. H. (1995). *Foundations of Databases,.*
3. Accenture; Deloitte Consulting; Gartner Research . (2001; 2001; 2002). *e-Government Leadership: Rhetoric vs. Reality- Closing the Gap; e-Government's Next Generation: Transforming the Government Enterprise through Costumer Service; The Gartner Framework for E-Government Strategy Assessment.*
4. AGIMO and IPAA. (2004). *Australian Government Information Management Office and Institute of Public Administration Australia (2004) Future Challenges of E-Government, Volumes 1 and 2.* Canberra.
5. Ahn and Bretschneider . (2011).
6. Ahn, M. J., & Bretschneider, S. (2011). *Politics of E-Government: E-Government and the Political Control of Bureaucracy.*
7. Andersen, K. N., Henriksen, H. Z., Medaglia, R., Danziger, J. N., Sannarnes, M. K., & Enemærke, M. (2010). Fads and facts of e-government: A review of impacts of e-government (2003–2009). *International Journal of Public Administration 33.*
8. Arellano-Gault, D. (2012). The Evaluation of Performance in the Mexican Federal Government: A Study of the Monitoring Agencies' Modernization Process. *Public Administration Review 72.*

9. ASSOCHAM. (2015). *E-governance and Digital India.*
10. Atzeni, P. D. (1993). *Relational Database Theory.* California.
11. Bangemann, M. (1997). The Information Society–the EU Framework. *Business Strategy Review*, pp. 13–16.
12. Batini, C. C. (1992). *Conceptual Database Design, an Entity-Relationship Approach.* Menlo Park, California.
13. Batini, C. C. (1992). *Conceptual Database Design, an EntityRelationship Relationship Approach,.* Menlo Park, California.
14. Bélanger, F., & Carter, L. (2012). "Digitizing Government Interactions with Constituents: An Historical Review of E-Government Research in Information Systems. *Journal of the Association for Information Systems 13.*
15. Brickley, D. (2015). *SWAD-Europe: Project Fact Sheet.*
16. Cannan, S. J. (1992). *SQL – The Standard Handbook.* New York.
17. Ceri, S. (1984). *Methodology and Tools for Database Design.* Amsterdam,.
18. Ceri, S. G. (1989). *Logic Programming and Data Bases.* Berlin.
19. Chamberlin, D. D. (1974). *'sequel: A Structured English Query Language.*
20. Chamberlin, D. D. (1976). *Sequel 2: A Unified Approach to Data Definition, Manipulation, and Control.* 1976.
21. Codd, E. (1971). *A Relational Model for Large Shared Data Banks.*
22. Codd, E. F. (1970). *A Relational Model for Large Shared Data Banks.*
23. Codd, E. F. (1970). A Relational Model for Large Shared Data Banks. *Communications of the ACM*, (pp. 377-387).
24. Codd, E. F. (1982). *Relational Database: A Practical Foundation for Productivity.*
25. Commission of the European People group . (1993).
26. Darker, Thompson and Fishenden . (2014).
27. Date, C. J. (1993). *A Guide to the SQL Standard.*
28. Date, C. J. (n.d.). *A Guide to the SQL Standard.* 1993.

29. Dawes, S. s. (2009). *Governance in the digital age: A research and action framework for an uncertain future.*
30. Division of Trade. (1993).
31. ECDL Foundation. (2008). *What is the i2010 Initiative?*
32. ElMasri, R. A. (1994). *Fundamentals of Database Systems.* Menlo Park, California.
33. Etzoni. (1999). *Identification Cards in America.*
34. European Commission. (2010). *IDABC.*
35. European Commission. (2015b). *European eGovernment Action Plan 2011-2015.*
36. European Commission. (2015c). *ISA: Interoperability Solutions for European Public Administrations.*
37. European Union. (2009). Ministerial Declaration on E-Government. *5th Ministerial eGovernment Conference.* Malmö, Sweden.
38. Fairly, R. (1985). Software Engineering Concepts.
39. Forman, M. (2002). *E-Government Strategy: Implementing the President's Management Agenda for EGovernment.*
40. Garson, G. D. (2004). *The Promise of Digital Government.*
41. Graham, I. S. (1996). *HTML Sourcebook.* New York.
42. Greenspun, P. (1999). *Philip & Alex's Guide to Web Publishing.* California.
43. Halligan, J. (2004). *The Quasi-Autonomous Agency in an Ambiguous Environment:The Centrelink Case.*
44. Holzer, M. (2004). *"Restoring Trust in Government: The Potential of Digital Citizen Participaton.*
45. ISC. (2015). *Internet Domain Survey.*
46. Kernaghan, K. (May 2003). *Integrated Service Delivery – Beyond the Barriers.* Australia.
47. Kimball, R. (1996). *The Data Warehouse Toolkit: Practical Techniques for Building Dimensional Data Warehouses.* New York: John Wiley & Sons.
48. Kleinsteuber, H. J. (2012). *Der „Information Superhighway ": amerikanische Visionen und Erfahrungen. Wiesbaden: VS Verlag*

für Sozialwissenschaften.

49. Liu, Y., Chen, X., & Wang, X. (2010). *Evaluating Government Portal Websites in China.*
50. Lum, V. Y. (1979). Proceedings of the International Conference on Very Large Data Bases,. *1978 New Orleans Data Base Design Workshop Report* (pp. 328-339). Rio de Janeiro, Brazil: IEEE.
51. Maier, D. (1983). *The Theory of Relational Databases.* Potomac, Md.
52. Maier, D. (1984). *The Theory of Relational Databases.*
53. Management Advisory Committee. (2004). *Connecting Government: Whole of Government Responses to Australia's Priority Challenges.* Australia.
54. Mathews, J. T. (1997). Power Shift. *Foreign Affairs* , 50-66.
55. Melton, J. (1996). SQL3 Update . *Proceedings of the IEEE International Conference on Data Engineering*, (pp. 566-672).
56. Melton, J. S. (1993). Understanding the New SQL.
57. Milakovich, M. E. (2012). *Digital Governance: New Technologies for Improving Public Service and Participation.*
58. Millar, L. (December 2004). *Networking government: e-government in New Zealand* .
59. Moore, M. H. (2000). *Creating Public Value: Strategic Management in Government.*
60. Moore, T. a. (2004). *The changing role of multi-channel service delivery.*
61. MOPAS. (2008). *2008 White paper of Public Administration and Security.*
62. MOPAS. (2010a). *2009 White paper of Public Administration and Security.*
63. MOPAS. (2011). *Smart Gov Plan.*
64. NAO. (2002). *Better service through e-government : Case studies in Support of Better Public Services through egovernment.*
65. National Information Society Agency. (2012). *"e-Government of Korea: Best Practices.*
66. NPR. (1997). *Access America.*

67. O'Neil, P. (1994). *Database: Principles, Programming, Performance.* San Mateo, California.
68. O'Neil, P. (1994). *Database: Principles, Programming, Performance.* San Antonio, California.
69. OECD. (2003). *The E-Government Imperative.* Paris: OECD Publisher.
70. OECD. (2003). *The e-Government Imperative.*
71. OECD. (2009). *Government at a glance 2009.* Paris: OECD Publishing.
72. OECD. (2011). *Government at a glance.* Paris: OECD Publishing.
73. OECD. (2012). *Government at a glance.* OECD Publishing.
74. OMB. (2001). *E-Government Strategy.*
75. OMB. (2012). *Report to Congress on the Benefits of the President's E-Government.*
76. Paolo, A., Stefano, C., Stefano, P., & Riccardo, T. (1999). *Database Systems concepts, languages and architectures.* Milan.
77. Paredaens, J. D. (1989). *The Structure of the Relational Database Model.* New York.
78. Paredaens, J. D. (1990). *The Structure of the Relational Database.* Berlin.
79. PCGID. (2003a). *Participatory Government Vision and Direction of e-Government.*
80. PCGID. (2003b). *e-Government Roadmap of Participatory Government.*
81. Pew Institute. (2010). *An Analysis of Continuity Intentions of e-Government Portal Users.* Public Management Review.
82. PORRÚA, M. A. (2013). *E-Government in Latin America.*
83. Ramakrishnan, R. (1997). *Database Management Systems.* New York.
84. Reece, B. (2006). E-Government Literature Review. *Journal of E-Government 3*, pp. 69–110.
85. Rosenbach, M. (2012). *Von der Agenda for Action zum Telecommunications Act von 1996: USKommunikationspolitik zwischen Deregulation und öffentlichem Interesse.*

86. Seifert, J. W., & McLoughlin, G. J. (2007). *State E-Government Strategies.*
87. Senn, J. A. (1989). *Analysis & Design of Information Systems.* New York.
88. Silberschatz, A. K. (1996). *Database System Concepts.* New York.
89. Sutter. (2010).
90. Teorey, T. J. (1982). *Design of Database Structures.*
91. Teorey, T. J. (1986). A Logical Design Methodology for Relational Databases Using the Extended Entity-Relational Approach. *ACM Computing Surveys*, (pp. 201-260). Englewood Cliffs, N.J.
92. Teorey, T. J. (1990). *Database Modeling and Design: the E-R Approach.* San Mateo, California.
93. Tsichritzis, D. L. (1982). *Data Models.* Prentice Hall, Englewood Cliffs, N.J.
94. Ullman, J. D. (1988). *Principles of Database and Knowledge Base Systems.*
95. Ullman, J. D. (1988). *Systems, Principles of Database and Knowledge Base.* Potomac, Md.
96. United Nations. (2005). *Civil Registration in Austria.*
97. UNITED NATIONS. (2014). *E-GOVERNMENT SURVEY: E-GOVERNMENT FOR THE FUTURE WE WANT.*
98. United Nations. (2014). *United Nations E-Government Survey 2014.*
99. Valerie Frissen, J. M. (2007). *The Future of eGovernment.*
100. Veit, D., & Huntgeburth, J. (2014). *Foundations of Digital Government: Leading and Managing in the Digital Era.*
101. Welch, E. W., Hinnant, C. C., & Moon, J. M. (2005). *Linking Citizen Satisfaction with EGovernment and Trust in Government.*
102. West, J. (1997). *Building Japan's Information Superhighway.*

INDEX